AF663731

BOOKS BY

BIJON KARMAKAR

The "Supreme - Universe" is the Great-grandmother of All Genesis.

Part 2

("The constructional aspects of the Baby Universe.")

Title of the Book:
The "Supreme – Universe" is the Great-grandmother of All Genesis, Part 2.
Subtitle: The constructional aspects of the Baby Universe.

First Published:
19th October 2024

Edition:
This edition was published by the Author: Bijon Karmakar

ISBN:
978-93-341-4068-2

Copyright:
© Bijon Karmakar, 19th October 2024

All Rights Reserved:
No part of this book may be reproduced, distributed, or transmitted in any form or by any means, including photocopying, recording, or other electronic or mechanical methods, without the prior written permission of the publisher, except in the case of brief quotations embodied in critical reviews and certain other non-commercial uses permitted by copyright law. For permission requests, write to the publisher, addressed "Attention: Permissions Coordinator," at the address below.

Dedication:
For all who seek to understand the vast Universe and our place within it! All the book series delves into the realms of science, philosophy, and myth, offering comprehensive insights into the intricate fabric of existence.

Contact Information:
For inquiries, contact: [author.bijon@gmail.com]

"List of books published by the author:"

1. *The "Supreme - Universe" is the Great-grandmother of All Genesis, Part 2: The Constructional Aspects of the Baby Universe.* Published on 22nd Oct 2024.

2. *The "Supreme - Universe" is the Great-grandmother of All Genesis, Part 1: Inception of the Actual Universe.*
Published on 7th Oct 2024.

3. *Energy and The Complete Universe, Part 4: Energy and Quanta Relation (Thesis & Analysis), GRAVITY: In Search of The Reality of Life!*
Published on 1st January 2024.

4. *Energy and The Complete Universe, Part 3: Energy and Quanta Relation (Thesis & Analysis).*
Published on 2nd Dec 2023.

5. *Energy and The Complete Universe, Part 2.*
Published on 22nd May 2023.

6. *Energy and The Complete Universe, Part 1.*
Published on 5th May 2023.

7. *MYTHOLOGY, PHILOSOPHY, AND SCIENCE: Book For the New Generations, Volume 2.*
Published on 16th Aug 2022.

8. *MYTHOLOGY, PHILOSOPHY, AND SCIENCE: Book For the New Generations, Volume 1.*
Published on 15th May 2022.

"Here are some hints about the key insights from the book."

This part of the book series, The Supreme - Universe is the Great-grandmother of All Genesis, offers an in-depth exploration of critical concepts in the formation and development of the Universe. It begins by providing technical clarity on the famous equation, E = mc^2 explaining its role in expressing the eternal and unbreakable relationship between the actual Universe and the Supreme - Universe.

The book then traces several significant technical milestones during the embryonic stages of the Universe's development, delving into the concept of time and its intricate technicalities. As it progressively advances, the narrative moves toward understanding the distinctiveness of the 1st-energy-belt and the steadiness of energy-barriers, attributed to the flexibility of the constituted energies or unit-value-energies.

Wherever the concepts are complex, the book suggests using appropriate drawings, such as those illustrating the Universe-Time-Clock (Ütc). These visual aids help explain the key technical points, like the idea of "absoluteness," which is comprehensible standardly through parametric statistics and extreme values (like ∞). The book also explains the construction of the Universal-Time-Clock (Utc).

Additionally, it provides a simplified cross-sectional view of the actual Universe and its three essential energy belts, which developed in harmony with the momentary nature of energy and time on the 'Universe-Time-Clock.'

This part of the book series discussed energy levels and constructional aspects of the 1st-energy-belt, providing technical clarity. Additionally, this section delves into the technical aspects of energy levels within the Supreme - Universe, including pivotal concepts like the Lymph-bubble and Lymphatic-energy in the Influencing Energy Zone

(IEsu). A comprehensive analysis of other significant aspects follows, including helicity-energies, energy strings, and the roles of Generic-energy and Genomic-matter. The initial phases of energy, some marked by the 'Nullifying force of energy physics,' are identified as the impactful actions in the foundation and structuring of the actual Universe.

Further chapters explore the Bullwhip effect in forming advanced energy components and widen the constructions of the Gestation Line, which gradually transforms and plays a crucial role in the development of the Universe. The book also critically analyzes the concept of 'equilibrium,' highlighting its genetic or inherent origin in the Pure energies of the Supreme Universe.

The book concludes with discussions on differential values, symmetry, asymmetry, and the energy atmosphere, all of which play pivotal roles in the gestation processes of the Universe.

Through logical explanations based on the standard 3D assessment practices of intelligent beings, the book emphasizes that differences in unit-value energies in their Pure state are the fundamental origins of everything, including the formation of the actual Universe.

Additionally, this part delves into topics such as Gravity, Waves, Atmosphere, and other subjects that stem from materialistic knowledge, enhancing living beings' perceptible understanding. All these phenomena have direct or indirect connections to energy and its momentary functions.

In summary, this part of the book series provides a comprehensive technical and philosophical framework for understanding the Universe's formation, structured energy levels, and the fundamental workings of energy elements that drive its continuous progression.

Index

Introduction

A short introduction of part 1 of the book series "The Supreme - Universe is the Great-grandmother of All Genesis."

Herewith, a drawing SU1 is a three-dimensional sight, merely to give an idea and convenience to the standard mindset of dimensionally restricted living beings about the preparatory stages' structural position (*well advanced of the embryonic stage*) of the baby Universe in the womb of the mother Supreme - Universe.

Continuance and presenting the summary of the book, "The Supreme - Universe" is the Great-grandmother of All Genesis, Part 1," where Philo-Scientifically clarified the inception stage of the actual Universe with many significant points like the Supreme – Universe is a state-of-Absoluteness, a true Singularity in feature containing Pure-energies at their Absoluteness-status, bearing all proprietorially distinguishable or indistinguishable diversifying working essences or energy elements that are constitutionally holding by the actual Universe upon getting the grandeur as an individual position.

The medical term 'infection' is conveniently utilized instead of applying a new expression to indicate the weakening of the immune system from the Pure -status of energies in some specific areas like 'IEsu' (*influencing energy zone*) that lead to inception in the vast state-of-Absoluteness of the Supreme – Universe.

But technically, the consequences of sequential and non-sequential sliding - motions gradually begin to induce and initiate deposition *(this deposition is the inequitable Motional values from the standardness)* at a point of unit-value Pure-energies and then slowly increase to reach a peak where Motional-values deposition became capable of initiating diversity, and then infecting the closer units of Pure-energies, besides infectiously spreading due to multidimensional effects to an infinitely larger area/s to initiate inception as an expectant mother for giving the birth of the actual Universe.

In addition to the above, many other techno-philosophical points elucidated in the book "The Supreme - Universe is the Great-grandmother of All Genesis, Part 1," concerning progesterone of the Great-grandmother Supreme - Universe and consequential diversifying roles of energy-Physics or elemental-values functionalities that led to the formation or inception of the actual Universe.

▶ Some technical clarity of 'E = mc^2' is appropriate to express the eternal and unbreakable constructional relation of the actual Universe and the mother Supreme - Universe (*presented with the drawing SU2*).

Honoring the fundamentality of equation E = mc^2, it may widen for convenience to understand the functionality and principles of the **'objectually profound distinctional infinite'** to **'energy levels non-distinctive infinite'** like state-of-Absoluteness of the Supreme - Universe and its subspace or subset-value like the actual Universe, which progressively consolidated to a unified law occupying an eternal position within the objectually presented limitless /unknown expanse, symbolizing the essence of perimetral energy levels cosmical existence.

*(Here, the **distinctional infinite** indicates where judgments are mainly determined through the physically functional nothingness following the constructional purview of intelligent species in this habitable globe. And, energy levels **non-distinctive infinite** signifies Truly a non-distinctiveness, which is applicable beyond any aspect of objectual and limits of non-objectual or simply the standard sense of diverse levels of energies in the matured 2nd-energy-belt.)*

▶ Following the drawing SU2,

If: -

a. The alphabet 'E' justifies representing the 'state-of-Absoluteness' of the Supreme - Universe, then the 'mc^2' can apply to the preparatory Universe or baby Universe considering the **'c'** an overall motion deriving only one type (*which is proprietorially an exclusive value*) of comparable thinner grades of energy that rules over the rest of thicker grades of energies the **'m'** that covers the area assuming as <2> or applying objectual sense, this <2> can replace by **<n>** (*somewhat an infinite position*) in fecund Universe.

(Here to note, for the Supreme - Universe, the standard expression through the word sense thinner grade of energy can be justified as motion-element ***(c)*** *at their truly the inherent flawless extreme to the lowest neutral Motional property, whereas the alphabet* ***(m)*** *is doable of utilizing to indicate the Pure-energy represent the true singularity, which is entirely unlike the earthly way presented uniqueness on some dimensional aspects.*

Also, the sense '<***n***>*' is applied to indicate some transcendent value that seems to be a cultivated application, but its infinity or endlessness performs beyond the limits of objectual senses that sweeten for broader understanding from the standard digital value '2.')*

b. Likewise, as a whole, upon maturity of the Universe, the equation 'E = mc^2' can be applied in a disintegrating manner to justify like the 3rd-energy-belt or speediest motion (*or may be considered as free-flow motion*) deriving 'Purest-energy' represented by the alphabet 'c.'

Then, the rest of the energies overall are contained in the 1st-energy-belt, and the 2nd-energy-belt represents the place of (m) for total coverage ^<n> of the actual Universe indicated by the alphabet 'E.'

(Here, the sense of total vastness coverage ^<n> technically may be justified on multidimensional aspects.)

c. In the same way, E = mc^2 can apply for the 1st-energy-belt, considering energies in the 2nd-barrier represented by the alphabet 'c.' Whereas functionally, the overall value of the 1st-shield and 3rd-shielding barrier act as 'm' in the total coverage ^<2> assuming surrounded by an alphabet 'E,' which justified as the total configuration for the 1st-energy-belt.

Also, other than the 1st-shield and 3rd-shielding barrier for the 2nd-barrier of the 1st-energy-belt, this 'E' is covered by the vastness of the Supreme - Universe, and internally the 2nd-energy-belt and 3rd-energy-belt, which indicated in the drawing SU3.

d. Also, the equation E = mc^2 can apply for the 2nd-energy-belt considering the ostensive Purest-energies as the motion-deriving component, signified by the alphabet 'c' and the rest of its overall containing energies represent as 'm' in the whole coverage symbolizing as ^<2> of the 2nd-energy-belt that surrounded by the 1st-energy-belt and the 3rd-energy-belt, which proprietorially represent as 'E' for the overall containing in the matured 2nd-energy-belt of the Universe.

▶ Further, the containing energies in the 2nd-energy-belt maintained relative relations whenever the energies took the values or assignments as mass-energy units, mass-energy frames, or objects containing motion-deriving 'c' comparably thinner than the rest of mass-energy 'm' in the area covers as ^<2>, peripherally regarded by the alphabet 'E' which is surrounded by the **atmosphere** regardless of smaller and bigger dimensional articulation.

*(Here, the word sense **atmosphere** justified as a somewhat **free-flowing** 'energy-atmosphere and Twig-atmosphere' mostly participated in conveying physics.)*

▶The randomness of the equation E = mc^2.

Based on the above expositions, it is reasonable to state that from the perspective of 3D discernment of living beings, the equation E=mc^2 proprietorially portrays randomness when judgment or measurements accomplish following some parametrically limiting constants, justifying the value of infinite (∞) as an inaccessible vacant, vacuum or nothingness that all functionally acts 0-like application to the logic of standard consciousness of human species and their created intelligent aids or like AI.

→Further, for clarity, the symbolic (E) is uniformly applied for energy to indicate a constant value representing a capital state font created by linguistically advanced intelligent human beings. But the small state font alphabet (m) and (c) indicate distinctive types of energies on the aspects of quality that have meaning where it picked randomly from the 26 alphabetic orders from (a) to (z).

However, the symbol ^ in the equation applied to indicate a 'power sign' and the number '2,' beside the (=) sign are far different picks from each other, and even the linguistically recognizes 26 alphabetic fonts.

Psychologically, all those different picks to make an intellectual equation like (E = mc^2) typically be accepted as a natural attitude or celestial gift, even characteristically all abide by random properties that consolidated into a simple unified one to indicate interchangeable relations with everything that performs unimaginably formulating functionalities as energy and its outgrowth as object forms in entire visible and invisible cosmos or the 2nd-energy-belt.

e. The 3rd-energy-belt contains only one type of energy, an extremely refined or thinnest form honored with the 'Purest-energy,' proprietorially the **True-singular** status in all aspects of functionalities since the Purest-energy truly is the extremity of kineticism, neutralism besides on dimensional aspect it represents truly an Absolute-zero for entire matured Universe.

*(Here, the **True-singular** status means, proprietorially, actually a unique and extraordinarily single or mono status, which is entirely unlike the unified status singularity of Pure-energies.)*

So, the 3rd-energy belt, as a single, can be justified as 'c' to portray the rest of the energies 'm,' which contributed to giving the position of the 2nd-energy-belt and the 1stenergy-belt for the dignity as the actual Universe that may dimensionally regard as ^2 since periphery surrounds by the 'E,' which is the state-of-Absoluteness of the Supreme - Universe, an eternal constant maintained evenhanded *(that is, =)* relation with the actual Universe contained with mass (m) and motion regulating (c).

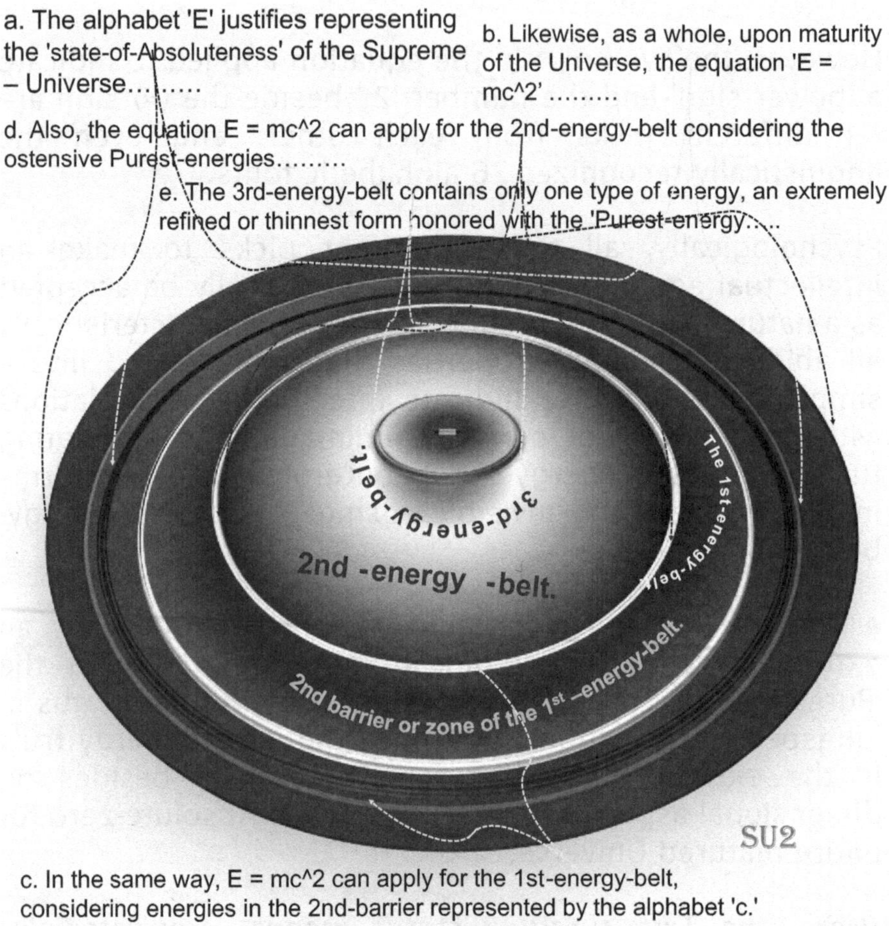

So, it is understandable that the fundamental equation E = mc^2 or enhanced value mc^<2>, or mc^<∞> is appropriate to present and define the actual Universe and its hidden secrets that are beyond the perceptible sense to human species, besides their intellectual new-new creations for aiding purposes.

▶ Part 2 of the book series, "The Supreme - Universe is the Great-grandmother of All Genesis," will give some more key points about energy level technicalities that acted in gross for the advancement of the baby Universe.

Also, to ease understanding through the standardness of intelligent living beings, it attempted to explain comparably, applying the reasonable word senses that are common and accepted by the contemporary generation to know about finite features of some states of matter or objects.

Continuance of earlier part of the book series.

Chapter 9

Here are some significant technical milestones during the gradual development of the embryonic stage of the Universe.

Following progression and gradual multiplication of differences, the intramural units of energies *(or affected Pure-energies that participated in the preparation of the Universe)* uniquely process due to divestment and inducement of constituent energy-elements that consequences of evolving with the new applications of the elemental - values (*or, these elemental values may be dignified earthly ways as physics values*) during gestation for the baby-Universe.

1. Time and some of its in-depth technicality.

Based on this earthly physics, it may speak that the energy times, which derive while Motional-values of motion -

elements of the units of energies became considerably unstable from the reasonable resting to celerity.

→So, consequential unstableness from resting or naturally speak, from the inert /cipher states **Timeless** to effective Time-emanating unit-value energies begin to disperse or consider as dissipate away from the downsized center of the **gestational** point that gradually or slowly infectiously widen the diversifying functionalities inside the 'IEsu' zone of the Supreme - Universe (this **gestational** point attempted to clarify visually *in the drawing SU12 of Part 1 of this book series*).

Or, very simply based on the objectual way the above may speak as the acceleration of restlessness sequentially begins to **cultivate** embryonic areas of the prospective Universe.

*(Here, the word sense '**cultivate**' may be justified on technical terms for formulating energy physics that rigorously and multifariously acted to make productive of the barren or inert state potential Pure-energies in the IEsu of the Supreme - Universe.)*

Further, following the objectual sense and analogizing from the perspective of energy, the objectual word sense '**timeless**' does not have any practical /true meaning since intelligent beings consolidated all undiscernible or unacceptable time to zero-values for the convenience of measurements during a span of existence, to an individual or group purpose following the specificity of structural value in a place of habitat and travel.

But the energy-time retains forever as an integral part of the motion element to any state of unit-value energy rooming in the 2nd-energy-belt.

(Here to note, at energy levels, any form of computable or incomputable time, like standard clocks' time and many

other objectual numerable and indeterminable times, are directly or indirectly linked with the deep-down elemental-time of energy element, integrated with unit-value energies regardless of they represent as a simple mass-energy or referable mass-energy frame.)

→Persistency of effective and ineffective "Motional - value" of motion element is one of the most fundamental operational properties of each discrete unit-value energy, whatever perceptible sense disrespect to accept following the limits of specific way designed intelligent species in any habitable nature or globe.

So, the word sense 'tenaciousness' is suitable to explain energy-time that became the primary property of each unit-value energy. Irrespective of the unit value representing an Absoluteness-status for the "Pure-energies" of the Supreme - Universe where the sense of 'time' reached the lowest of incomprehensible and indivisible property of Absoluteness-status, which are unlike the cultivable times integrated with units of energies occupied by birth in the actual Universe, and in progression that carried on even after the maturity of the Universe.

Accordingly, the reality of the sense 'time' persists to any conceivable unit-value 'Pure-energy' in the Supreme - Universe and even to the optimum extreme, like a unit-value 'Purest- energy' of the 3rd-energy-belt, whatever the energy-times represent as imaginable active time or non-imaginable timelessness extreme stagnant to intellect senses of diverse adapting natures or globes.

→Therefore, it is reasonable to say about the eternalness of energy-time or represent it *(time)* as more in-depth of elemental-time since any form of discrete unit-value energy occupies some level to an extreme lowest or highest degree of motional-values effective or non-effective accretive or regressive assignment of motion-element.

2. Progressively step up towards the distinctiveness of the 1st-energy-belt.

Here, philosophically say, 'in the hope of some right expectation,' or the same technically present as, gradually toward the **easy-going approach** and development for maturation of the Universe, some energy times of the preparatory units of energies get sealed as a **dead-reckoning normal motion,** which is excluded in many aspects of functions and due to the lowest levels' energy physics.

Gradually in the process, those types (*means energy units with **dead-reckoning normal motion***) of preparatory energy units were getting organized to appear or surface as the energy levels' shield to uniquely sectioned like 1st-shielding, 3rd-shielding lines /zones, and some settled in the 2nd-barrier or zone, for preparation of the 1st-energy-belt within peripheral line of the Supreme - Universe and the immediate limits of the 2nd-energy-belt in the prospective actual Universe *(indicated in the drawing SU2).*

*{(Here, the word '**easy-going approach**' is justified to indicate a medical term concerning some easily and free-spreading infection.*

*And, the sense of '**dead-reckoning normal motion**' is somewhat (that is, not very perfect) equal to the non-functional or non-effective times, reasonable to apply as the sense of initial times at an 'Absoluteness-status of Pure-energies.'*

*So, it is suitable to say that the units of energy, which are at their 'dead-reckoning normal motion,' contributed to organizing the 1st-energy-belt in coordination with the rest of the disturbed energies in the IEfu or baby Universe, besides aiding from **functionally non-featuring** state-of-Absoluteness of the Supreme - Universe.)*

(Here, the word sense ***'functionally non-featuring'*** *is somewhat similar to the observably justified rigid value or assumed as a non-function line or bar, which justifies as a constant, base, foundation, or like 0-values' dimension comparably to analyze and indicate other dimensional activities.*

And, even the sense ***'functionally non-featuring'*** *may apply to human beings' 'mindset or predeterminate value' that mostly tends to act as a constant for analyzing and computing through the in-built physical system for the attainable and assessable items or imaginations.)}*

i. Further, referring to the sketch SU2 and in a stretching manner, the same in the above may attempt to clarify and say that some preparatory energies energy-times settled as a dead-reckoning regular motion between the peripheral boundaries of the Supreme - Universe and the prospective 2nd-energy-belt, which gradually in progression led to derive most of the energies in the generable 1st-energy-belt towards the formation of its 1st-shielding and 3rd-shielding zones to become **congeal** or **firm**, pushing the rest of energies or acting energy times to give the marking as the 2nd barrier, proprietorially a state-of-balancing ostensible absolute-zero status.

(Here, the objectually utilizing word sense ***congeal, firm,*** *or likewise seemingly be inappropriate, but it is suitable and easily convincible for comprehension instead of creating new words for reasonable expressions to define energy levels works.)*

(Technically to note, indeed, based on the delimited observable sense, it is to speak that lots of inexplicable energy levels of physics consistently enriched from the inception to become mature the actual Universe.

However, the exception is that upon maturity of the Universe, all energy levels generable physics or elemental values functionalities settled as a true constant or invariable status that denies the aspect of living beings' conception concerning any latest discovery on the point of energy.)}

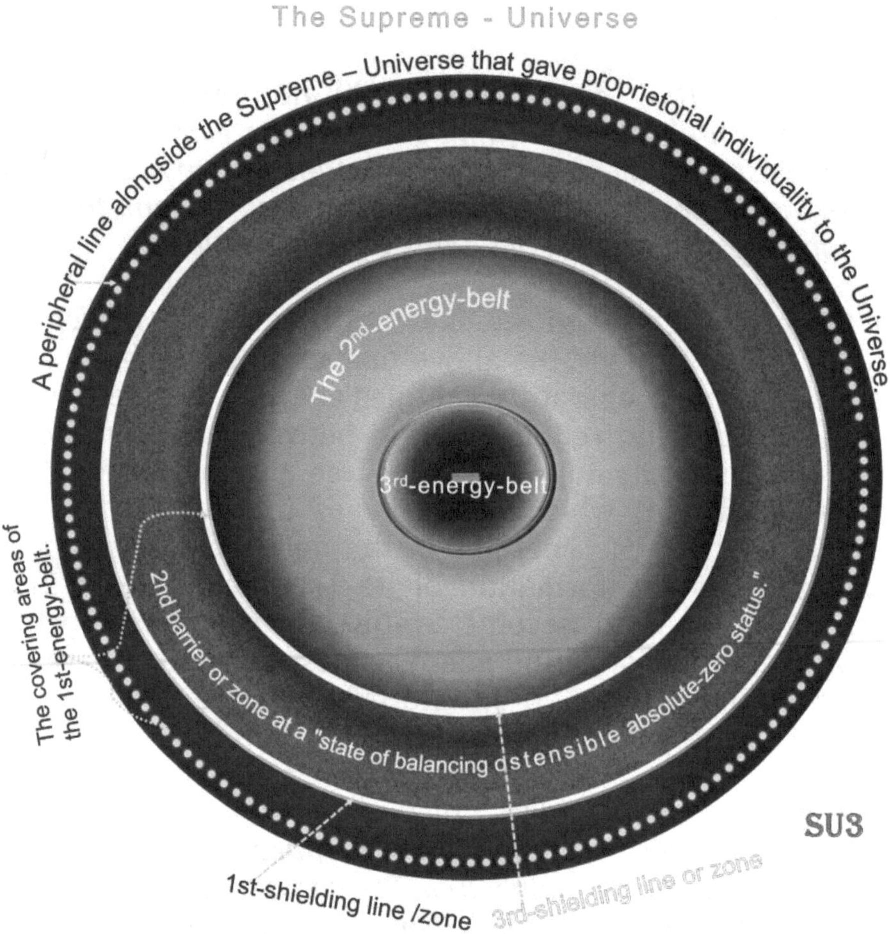

→It also may clarify that the 'energy times' of the preparatory units of energies that settled during the formation of the 2nd-barrier of the 1st-energy-belt is not at all 'ostensible absolute zero.' Instead, they are **variable /flexible** for maintaining balances if any contingent arises upon possible inducement of commotion or slumbering energy-times that are Motional-values or other elemental values effects from surroundings, as well as the progressing 2nd-energy-belt.

So, technically, may speak, those **variable /flexible** energies in the 2nd-barrier enable the raising of 'elemental values or elemental-physics' to maintain flexibilities in cooperation with the 1st-energy-shield and 3rd-energy-shield of the 1st-energy-belt.

→Or, with sketch SU3, which depicts a sectional view of drawing SU2, the above points may be clarified for ease of understanding in a little isolated manner by applying observable sense based on the constructional view of a standard time clock or wristwatch.

If the clocks encase *(here, 'encase' acts as a computational constant or rigid value)* beholds as non-accountable dead-reckoning-times for flexible movements of hour, minute, second hands, and so on to give the countable /regulative times, keeping the spindles as the base at a center position for indications of an external implicating source, then on three-dimensional outlook it will look like that all diversities generates in the 2nd-barrier restrained within it, due to 1st-shielding and 3rd-shielding lines /zones of the 1st-energy-belt.

→Referring to sketch SU3, a little more may clarify the above analogically and speak that the 1st and 3rd-shielding lines act as the clocks' cover when the 2nd-barrier performs the hands of the clock for the 1st-energy-belt, and during the movements of the hands of the clock indicates the flexible capacity of its times that regulate the motional-values effects and act upon due to the external implication.

Therefore, energy times from the preparatory units of energies that are settled in the 2nd-barrier /zone of the 1st-energy-belt, permanently resting at a balancing-ostensible-absolute-zero status.

It also may mean, gradually and progressively, the 2nd-barrier of the 1st-energy-belt became the acting life to secure the balance of the 1st-energy-belt, whatever rest of areas energy-physics impact on it (*1st-energy-belt).*

→In drawing SU4, the clock indicates countable /regulative times that are easily discernible on the three-dimensional judgment through the specific way evolved living beings, which is somewhat reasonable to understand the example of energy levels' steady multidimensional Time-regulating functions that determined the perfect secureness or balancedness of the 1st-energy-belt in the vastness energy-Ocean of the Universe.

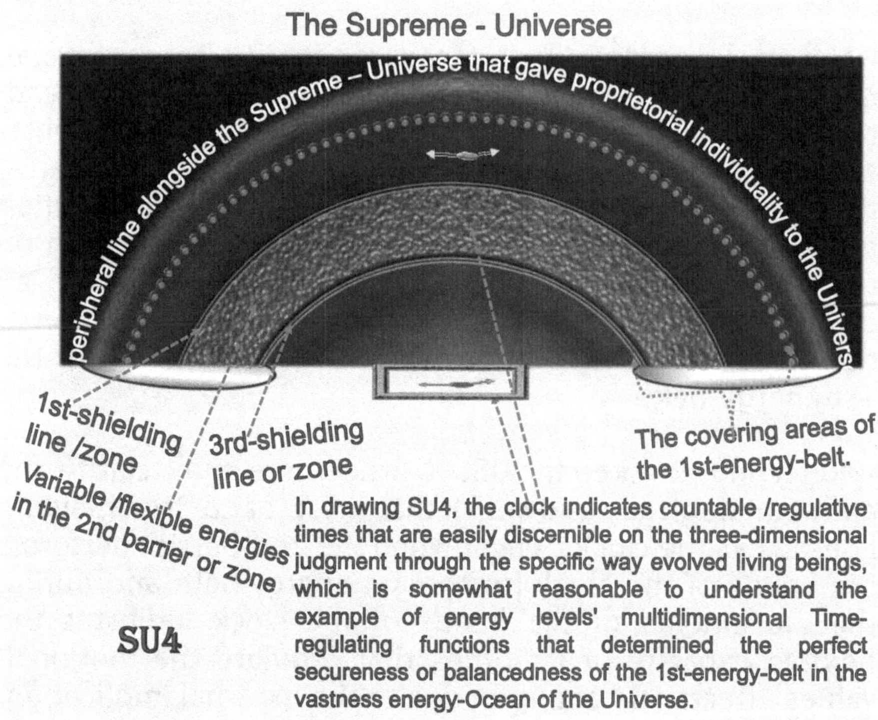

(One significant technical point to note is that the three-dimensionally (3D functionally means, 1D, -0D-, 1D) designed standard clock regulates to move in one direction, or always a positive move or 1D for easy flow of countable times following an origin, base point, or like 0-value (0-D)

while third-dimension of 3D or negative confined to obey undisturbedly the positive role of 1D. Accordingly, observable times became the mindset of living beings following the practicality of daily accessible items.)

ii. The steadiness of the 2nd-barrier is due to the typically flexible features for the constituted unit-value-energies.

Analytically, based on some unflinching types (*or degree-wise somewhat similar to pristine /undiluted property from Pure status*) of units of energies of the 1st-shielding-zone' and the '3rd-shielding-zone,' constitutionally sets energy - elements of the units of energies of the 2nd-barrier are composite with **devalued /dissolute** category.

And, from an operational viewpoint, the same *(that is, **devalued /dissolute** category of energies)* conveniently speaks that the units of energies in the 2nd-barrier conceived with **'redundant energy elements,'** so their elemental - values could efficiently run for deriving internal physics (*or internal 'elemental-physics'*) in the 2nd-barrier while in preparation for the 1st-energy-belt.

*{(Here, following the objectual sense of '**redundant energy elements,'** which technically indicate multi-proprietorial components or are assumed as energy elements that efficiently can handle multiple functions instantaneously upon any form of introduction or encountered externally, internally, or both countenances at a time.*

*Although, in practice and general, energy elements have the most high-versatility properties, thus any form of discrete unit-value energy somehow could manage their original dignity as 'energy' regardless of their **degree-wise adaptability**, which (means **degree-wise adaptability**) primarily applies to objectual senses to distinguish the properties of objects based on the quality, quantity and determining the outcome of recognizable physics (means,* *internal and external physics) to a microparticle or massive structural values object for denoting independent identity or nobility.*

Otherwise, the elementary level to any unit-value mass forms energy or speaks as energy elements in any mass-energy unit functionally versatile on the point of energy-physics since constitutionally set working components or energy elements capable of adjusting inconceivable momentarily, immediately upon inordinate application of external atmospheric physics.

This energy level adjustable capability is one of the inherent gifts obtained from the property of the Pure-energies while secured at extremely healthy status in the mother Supreme - Universe.)}

→Also, considering the flexible features of containing units of energy overall, the 2nd-barrier accomplishes to deter, check, or manage any external or internal swaying activities that may roll on the time-phased manner of energy-clocks of the 2nd-barrier and the 1st-energy-belt.

→Whatsoever, in short, due to constructive intuition and configuration with the growing process towards maturity of the Universe, energy-times of the 1st-energy-belt comprehensively reached almost like a state of individuality /uniqueness, or as a unit like **0** or complete **cipher-state** on the point of **functional conditioning**.

*(The objectual sense '**functional conditioning**' usually recognizes only some dimensional aspects, but on energy levels, it (*1st-energy-belt*) exercises multi-directionally.)*

{(Indeed, upon maturity of the Universe, all three energy belts reached their cipher individuality on the aspects of personality.)}

3. Some notes concerning the projection of Universe-time-clock = Ütc. *(Drawing SU5, SU6, SU7, and SU8)*

The expression or word sense 'time' is accountable even for the unimaginable most micro momentariness energy levels works, which is quite unlike the standard model of 3D conception that fundamentally applies *(directly and indirectly using natural or non-natural aids)* through the perceptible sense of intelligent beings in this biological nature.

So, following the diversifying energy levels roles from the initiation or birth of the Universe, the sense of times became a remarkably productive expression to convince systematic evolution of the Universe.

Accordingly, the 1st-energy-belt may project as a constructional checkpoint or like the 0 for the stable framing of the 'Universe-time-clock = Ütc.'

(Or, technically the same, that is, the 1st-energy-belt can justify as one of the standard frames or boundaries for a reasonable comprehension of 'times' that derived multidimensionally throughout energy levels, the most momentary works in the matured 2nd-energy-belt of the Universe.)

→On the way towards maturity of the Universe, its 1st-energy-belt gradually turned into a competent constant position somewhat similar to a standard time clocks' encase or frame of a wristwatch.

Or the same (*that is, the 1st-energy-belt*) may justify applying an energy levels end-point /boundary line to deter, block, or immobilize all forms of 'energy-times' that could yield from 'energy-time clocks, or energy-clocks' by the units of energies while in the process toward progression and even after maturation of the 2nd-energy-belt and 3rd-energy-belt.

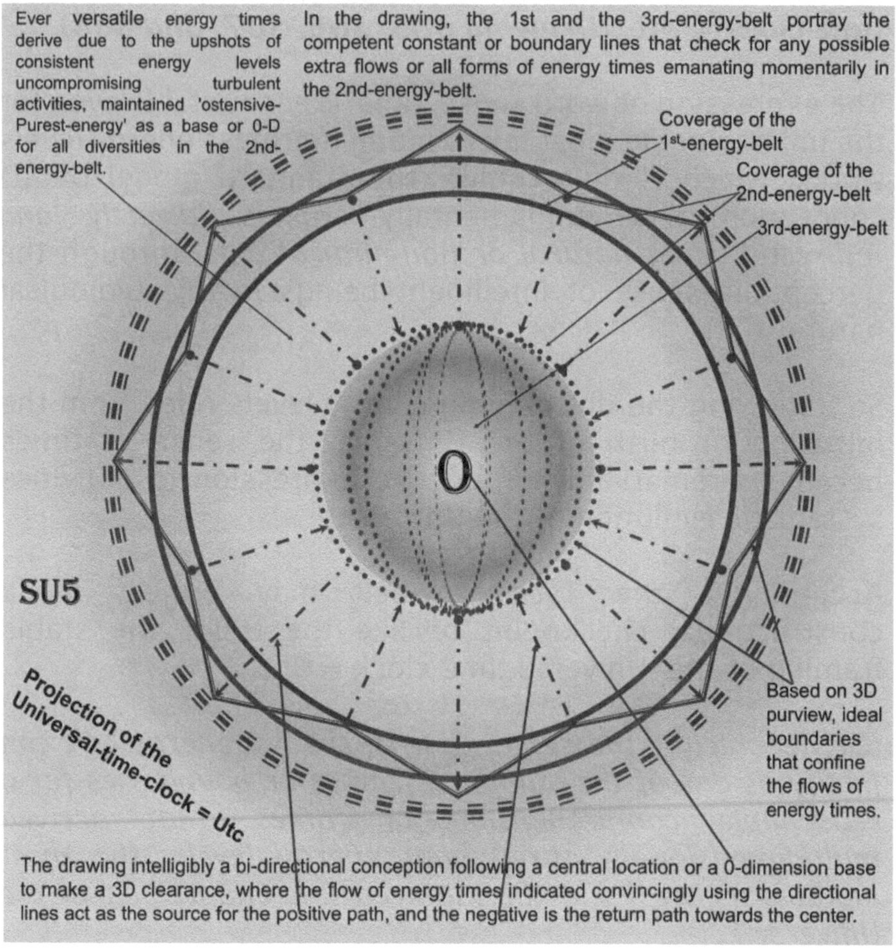

→Also, the effective energy times emanate by the units of mass form energy, mass-energy frames, objects, etc., while consistent with diversifying activities in the matured 2nd-energy-belt, both the 1st-energy-belt and the 3rd-energy-belt act as a two-dimensional (2D) boundary for derivable multi-directional energy-times in the 2nd-energy-belt.

{(However, analogously as displayed in the drawing SU6, whenever justifying through the sense of 'observable times' following a standard table clock, the 3rd-energy-belt represents a center point while the clocks encase (means, energy levels 1st-energy-belt) *acts as a boundary of*

measurable movements derived during the momentary advancement indicated through the clock's hands.)}

This drawing SU6 indicates an analogous presentation whenever justifying through the sense of 'observable times' following a standard table clock, the 3rd-energy-belt represents a center point when the clocks encase (means, energy levels 1st-energy-belt) acts as a boundary of measurable movements derived during the momentary advancement indicated through the clock's hands.

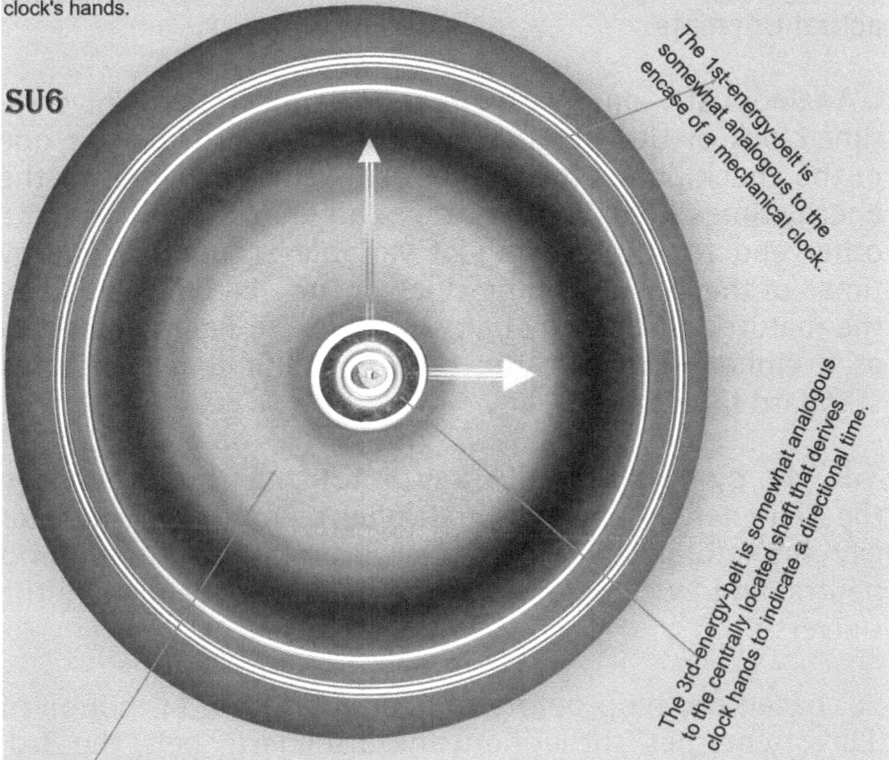

Ostensive Purest-energies of the 2nd-energy-belt independently act as a constant for the momentarily emanating *(without particularity of any direction)* energy times for the Universe-time-clock within the zero-reckoning times of the 1st and 3rd-energy-belt of the matured Universe.

→Little more may clarify the above and say that the 'energy-times' of any form of 'energy-clock' are due to motional-values works obtained by birth or speak as the inherent property of motion-element ingrained in each discrete unit-value energy of the Universe. Irrespective of types, grades, classes, or sub-class energies, besides the exclusively thinner form of energy, which represents a solitary type like ostensive-Purest-form available in the matured 2nd-energy-belt only.

Also, times persist to typical unions as mass state distinguishable or indistinguishable objects and objectual products, some of them in general, identify following the predefined class of some state-of-matter like solid, liquid, gaseous, plasma form that reasonably assessable by the intelligent beings in the matured 2nd-energy-belt of the actual Universe.

i. Analogously, on the bidirectional logic, the 'Universe-time-clock or Ütc' defended by the 1st-energy-belt as one of the check points for all motional-values activities or the energy-times that yield in the 2nd-energy-belt, and the other end for the ceasing of variable /countable energy times of the 'Universe-time-clock or Ütc' will be shielded by the matured 3rd-energy-belt since the 3rd-energy-belt acts as a cipher value like the central positioning shaft of a standard table clock.

→By the by here may give a little touch and quote that in the 3rd-energy-belt, motional-value of motion-element or say the motion of Purest-energies reached to uniform neutral and extreme peak on the point of dynamism in the Universe.

So, based on the property of a singular form of motion of 'Purest-energies' throughout the 3rd-energy-belt, the 3rd-energy-belt presumes a "**state-of-Absolute singular**" with wholeness feature with the extremely flimsy or purified /refined form of energy featuring the extreme-finest form of motion-element where elemental values or working essence of motion-elements reached to the one or single state to give unblemished neutral and the extremely peak on the point of dynamism.

*{By the way, here to mention that the word sense 'state-of-Absoluteness or Absoluteness-state,' which is the true **singularity**, applied only to the Supreme - Universe to discern the quality of energies, entirely unlike the **'state-of***

-***Absolute' singular*** *for Purest-energies in the 3rd-energy-belt.*

Also, it is to clarify that proprietorially on the point of energy, the word sense 'singular' and 'singularity' applied distinctively, one for 'Purest-energies' in the matured 3rd-energy-belt and the next one, that is, the singularity for the Pure-energies in the Supreme - Universe.

*So, from the mathematical perspective, both 'Pure-energies (*singularity*)' and 'Purest-energies (*singular*)' represent the extreme to infinite (^∞) position, where all forms of creative intellectual ideologies reach an unscalable* (fundamentally dimensional aspects*) true end.}*

→Therefore, on the point of three-dimensional (3D) practicality, the Universe-time-clock or 'Utc' obtained two checkpoints (*one is the 1st-energy-belt and the other is the centrally located 3rd-energy-belt*) for any form of energy-times resulting in elemental-values works, which may generate in the ever-turbulent 2nd-energy-belt of the matured Universe.

→ *{(One significant point is that when the matured 2nd-energy-belt is considered an overall independent 'energy-clock,' the constituent ostensive Purest-energy (*indicated in the drawing SU6*) functionally represents a centrally located motion regulating hegemony.)}*

→Here, it is to note that the sketches SU5 and SU7 for the 'Universal Time Clocks' are just specimens, provided to ease an overall bi-dimensional comprehension of the specific way-shaped intelligent species, which may not be accurately matched when analyzing the True multi-dimensional roles of energies, mainly for the most momentarily variable functions of energies in the 2nd-energy-belt.

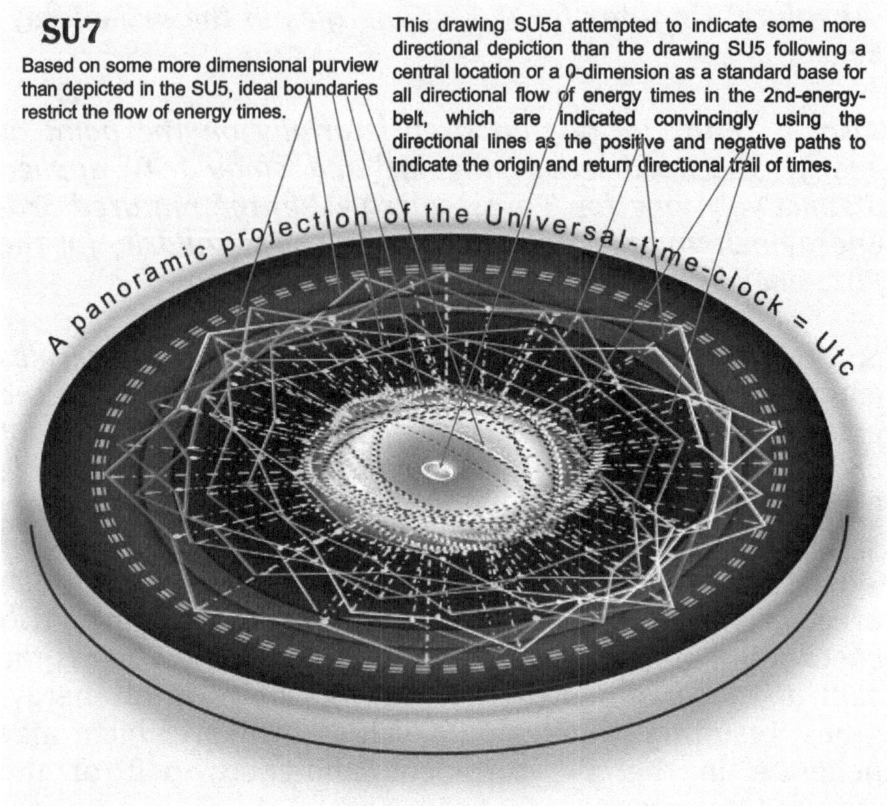

Also, especially for measurement purposes, it may speak that all units of living beings independently or Groupwise predominantly function as '0-D or cipher position' to analyze and accept permissible or reasonable variabilities in the surroundings, otherwise to indicate by mean of standard lines or curves, the energy levels directions and directional roles somewhat will act as entirely unacceptable jumbles scenario or image to the standard senses of human beings.

ii. Some notes on the objectually dominated word "absoluteness," which mainly determines following the standard sense of parametric-statistic of extreme peak (∞) or lowest (∞) values motion.

Shortly and comprehensively, the sense of the word absoluteness, in general, applied to the extent of certain easiness based on some objectual properties irrespective of extreme to the lowest and highest degree {*generally, mensurable degrees comfortably follow some distinctive 2D (+ and -) marks*} on typical or imaginative parameters.

→Therefore, concerning the work of motion elements, the parametrically evaluated sense of absoluteness is feasible to apply extendedly for the energies, following dynamical excellence that begins from the extreme lowest of degree /level and the extremity of the extreme peak of any intellectual presentation, which also may *mean on 3D purview,*

> "lowest^∞ -0 (*acting intelligent, or representing constant for measurement*)- peakest^∞."

→Factually, it is to note that the energy levels protean or ever-versatile 'motion' to the sense of living in any nature of the 2nd-energy-belt generally restrains to some extent and usually measures for reckoning on predetermined values.

So, following the standard meaning or objectual word sense as 'absoluteness or utter completeness' is inappropriate to apply for units of energies that became integral parts of the matured 2nd-energy-belt since any form of object and objectual product abides by the law of transformation.

iii. Remote in an analytical manner based on energy, the usual logic senses of intellect living, it may reflect that there is the possibility of numerable energy-times in an energy-time clock or speak as energy-clock.

Indeed, whenever and wherever the elemental activity of constituent energy - elements of unit-value-energy that attained a certain stable status, regardless of the energy units represent the appearance of objectual-energy, objectual-elements, objectual-products and so on advanced stage's objectual products, like matters and materials that all act as energy clocks, where the average energy times encompassed within the frame/s representing as discrete unit-value mass-energy, mass-energy frames.

→Whatsoever, accountability of 'energy times' integrated with the motion or while performing Motional roles due to constitutionally sets motion elements in any form of discrete unit-value energy or their composite state representing object/s.

However, it is to speak, immediately upon extreme and aberrant flexibility due to unusual application of physics, it (*means, unit-value energy or their frame as object*) may fall beyond the limit of the parametrical logic of living sense that can disallow getting the perfect values while imagining to measure following energy-time-clock or energy-clock.

Technically, computable energy-times of the Universe-time-clock, universe-time, energy-time-clock, and so on the variable motions, or say, as comparable and countable motions, play their persistent role within the limit of the matured 2nd-energy-belt since rest coverage like 3rd-energy-belt and 1st-energy-belt along with the energies in the shielding barrier of the matured Universe reached at a mensurable perpetual constant to 2nd-energy-belt.

Further to quote for ease of comprehension, upon maturity of the Universe, the 1st-energy-belt and 3rd-energy-belt independently reached their absolute uniformity at **'zero-reckoning energy time'** (*indicated in the drawing SU8*) on the aspect of the Universal-energy-clock.

(Or, the sense of ***zero-reckoning energy time'*** *may be considered an absolutely no derivable eternal yield from 1st and the 3rd-energy-belt, whatever the internal yield*

performs to sustained individuality by balancing from own derivations.)

iv. The drawing SU8 is a comprehensive display that endeavored to indicate some points on comparable strategies about the constructional aspect of Universal Time-clock or Utc.

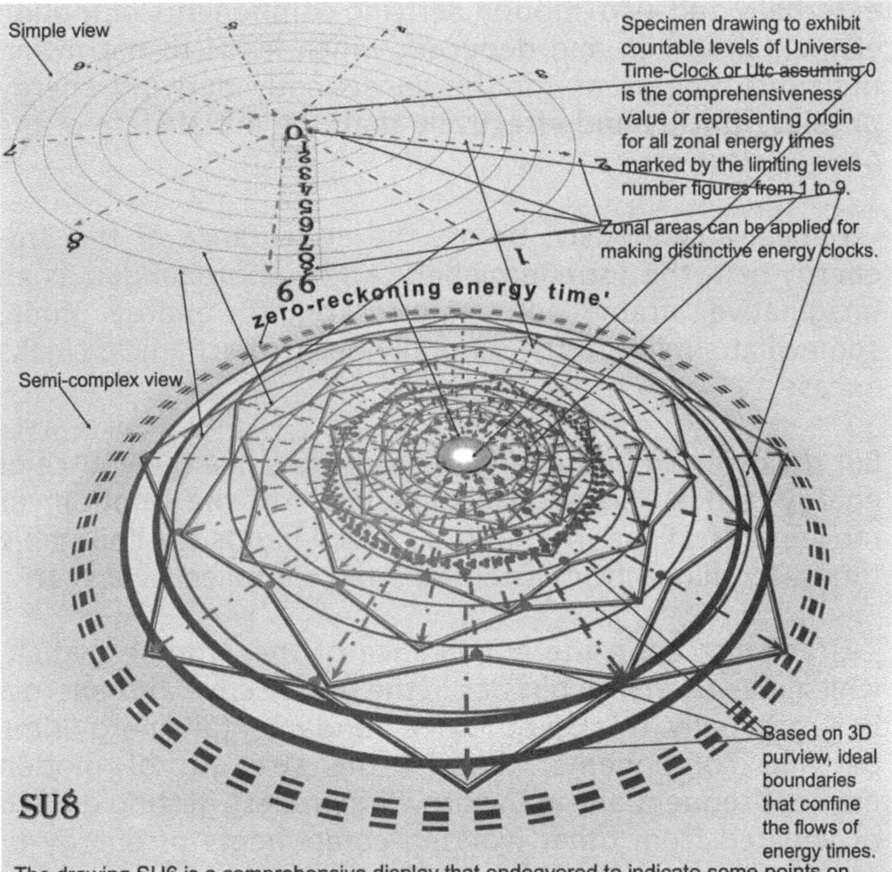

The drawing SU6 is a comprehensive display that endeavored to indicate some points on comparable strategies about the constructional aspect of Universal Time-clock or Utc.

Technically, in a brief manner, while in progress or before the formation of the 3rd-energy-belt, the prospective Universe-time-clock = Ütc, trapped or entangled and set about the growing periphery of the 1st-energy-belt, which is alongside the Supreme - Universe, so this resulted in functioning and organizing the energy levels physics or momentarily appearing elemental physics effectively within it the baby Universe.

Gradually, towards advancement and upon maturity, energy times of units of energies in the 1st-energy-belt comprehensibly stuck at the zero-reckoning energy time or perceptible no-motion zone since there is no effective yield externally and also, motion deriving components or motion elements weaken and deduced, which leads to narrowing the functional areas for the consistently deriving energy physics, quickly and effectively than the initial state of the growing Universe.

→Hence, analogously, before the appearance of the 3rd-energy-belt, the 1st-energy-belt can be apprehended as an imaginative frame of zero-reckoning energy time, somewhat similar to the naturally borne mechanical clocks encase.

But the growing 2nd-energy-belt remains like the source of energy times, somewhat alikeness of the motion or movements of the mechanical clocks' hands for countable times, regulated from a central source implied by a shaft.

→Gradually, upon the appearance of the 3rd-energy-belt, which means energy physics in the prospective 2nd-energy-belt gradually weakened since fundamental energy-time deriving components, that is, the strength of motion element sequentially in the growing process getting diluted or isolated from other working components of energy or energy elements.

Besides, consequential physics values centripetal functionalities lead to deriving the filtered or finest form of motion element toward the center for constructing the 3rd-energy-belt.

Briefly, consistency of progression and consequences of some particular stages of effectual functions, the 3rd-energy-belt finally reached an individual position by turning to a zero-reckoning energy time to give a genuine non-computable uniform role /activity for the Universal-time-clock.

(For the 3rd-energy-belt, the sense of zero-reckoning energy time is reasonable to speak as non-productive energy times.)

So, technically, the rest of the inner energies in the 2nd-energy-belt consistently play with versatile activities within the perimeter of the 1st-energy-belt and the 3rd-energy-belt of the matured Universe.

v. For ease and conceptional clearance, some points may speak about the standard practices obeyed by the human species in this bio-nature regarding countable times, which evaluate depending on the standardness of assessable variable movements of objects and objectual-products in reliance on some base position or presumptive constant in surroundings, including measurer.

It also may mean that, in general, all sorts of values of the moment and their follow-on variabilities are computable in a comparable manner when there is an accessible position as base or constant.

Like a mechanically structured 'time clock,' each movement of individual hands is countable by specific or recognizing units acknowledged by the word-sense as micro-second, millisecond, second, minute, hour, thereon as day, month, year, and so on distinguishable dignity /identity holding the observer and its inbuilt overall value of living-clock as an invariable or constant for reading those countable times.

So, it may speak that all those countable presentations are on comparable strategies maintained on some mensurable pre-determinately extents from the logic of lowest to highest levels. like: -

> ►<??..... <microsecond <millisecond < second < minute < hour < day < month < year <era, and so on, up to the standardness of pre-defined parameters.

→Technically, from the standpoint of energy, typically structured intellects of this bio-nature measure all sorts of assessable movements of objects and objectual-products depending on the standardness *(here, standardness means, logic acceptance)* of variable momentum caused by deeply established unsteady sway of motional value of motion elements.

→Thus, for future generations, this objectually defined word-sense time may convincingly and comfortably extend for the application /measurements for the acts of energies and their units and union as the smallest or largest form of mass-state or objects, irrespective of their presence in different Cosmical-Universes in the 2nd-energy-belt.

vi. Because of 'energy,' energy times are countable while there is an act of motion-element/s. Or the same may be defined as each momentarily variation of elemental values of motion element or overall momentum of unit-value-energy, or their admixes as a mass form unit object/s or objectual-products derived with specific values loss or gain of energy as ≜ on apprehending measurable scale/s.

(Here, the symbol '≜,' that is, delta over equals in mathematical terms used on the sense as a new variable or function.)

→Following the progressively appeared sense as 'standardness' of biologically shaped contemporary intelligent species, a little more may speak that all sorts of acts and effects of energy elements are possibly mensurable on comparable strategy in reliance on typicality of next generations inventing parameters for measurements of diverse energy levels functions in the 2nd-energy-belt.

The above is somewhat an improved intuition instead of the standard classical concept that determines the overall working outcome of units of advanced class objects and objectual products like some distinguishable states of matter features as solid, liquid, gaseous, and plasma that are standardly recognized to represent mass-energy frames.

In the mass-energy or mass-energy frames, regardless of their distinguishable or indistinguishable features, the constituent energy elements,

> ►like the act of motion-element that compels to derive variable motions,
> ►the act of temperature-element derived different degrees, featuring cold and hot temperatures,
> ►the act of pressure-element on their high and low intenseness,
> ►and so on to the infinite featuring energy elements that constitute the unit-value energy, mass-energy frame, or their advanced stages objects and other products.

{Indeed, shortly may touch and speak that classically defined physics (like movement, pressure, temperature, and so on expressed through books) *is the overall outcome of constitutionally set energy elements or working essences of any discrete unit-value energy whenever and wherever justifying the unit-value as matter or object representing a mass-energy frame.}*

→Further, technically on energy point, it may speak that the momentum of unit-value-energy or presuming as countable energy-times derive with changes /alternation in the course of devaluation or evaluation of energies with some form of loss or gain of the elemental values from the set-in energy-elements from certain standard status of unit-value energy, which displays as an overall internal outcome for an observable object or the representing mass-energy frame.

4. The drawing SU9 is a simple projection of a cross-sectional view of the actual Universe and its prospective 1st-energy-belt that accompanied the other constituent sections. Which sequentially developed in the harmony of the energy-times of the Universe-Time-clock of the Universe.

Convincingly say that the physics values or energy levels elemental values applications, following the Universe-time-clock led to organizing properly the 1st-energy-belt and its different protective layers or sections.

Technically, it happened due to the time-based consistent evaluation and devaluation of energies that allow to break off and reorient in a repeated manner the diverse form of energy elements that once were set as singularity states in the units of Pure energies in the Supreme - Universe.

→For ease of comprehension, at the early stage concerning the developing Universe, a little more may delineate from the drawing SU9 that each momentariness of motional-value of motion-element, or say as energy-times consequences with some form of energy loss or gain, those have participated actively during the preparation and structuring the actual Universe.

A cross-sectional view of the actual Universe.

SU9

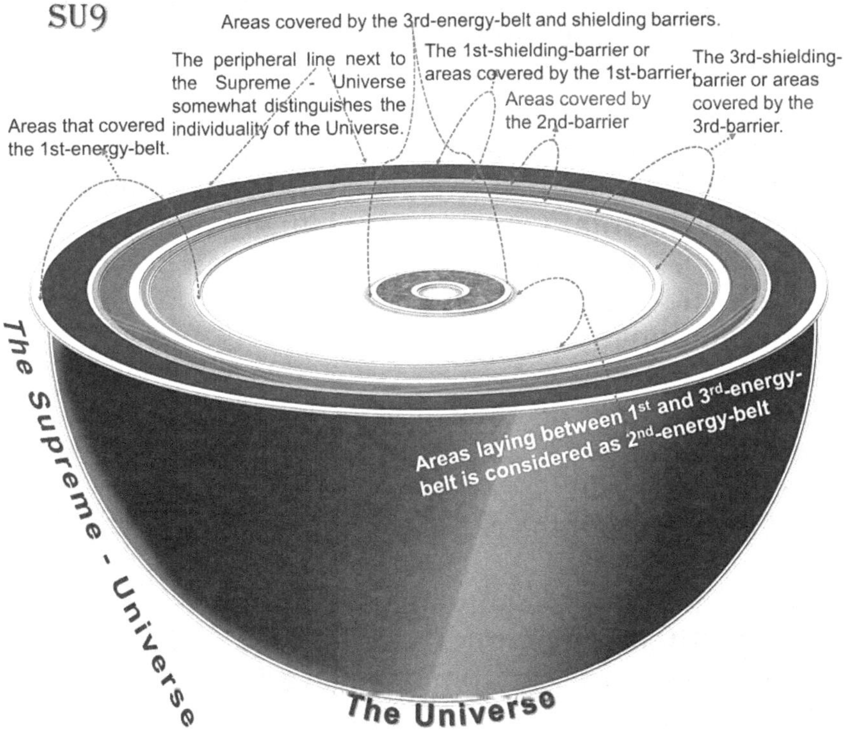

(From the above, one thing is clear to comprehend the momentariness of 'Motional values' at the elementary level of energy elements or objectively the most 'ground level' or utilizing sense as the 'root cause' of deriving times.)

Consequences of loss and gain of energies from assuming an earlier stable position of any presumptive unit-value mass-state energy emanate from types of constitutionally developed elemental-values effects.

Or the same, simple objectual way may speak as the physics-values effects derive due to energy levels constitutionally set working components or energy elements.

→For instance, consider that the temperature - elements and the motion - elements are only the two of many **bewilderingly unified**, integral parts of a unit-value alloyed or impure energy, and when those two proprietorially distinctive energies - elements (*like the temperature-deriving temperature elements and motion-deriving motion elements*) mixedly release their effectual elemental values or objectually depict physics - value. Then, its effective outcome acts as an objectual way of defining physics - values that may present as **Thermodynamic** work.

{(Here, the sense of ***'bewilderingly unified'*** *is utilized plainly to convince following the limits and existing perceptions of intelligent species that are specialized to analyze and accept the reality on some directional aspects.)*

Further, following contemporary science, expression through **'thermodynamics'** standardly represents the transfer of heat energy from one place to another and from one form to another.

So, directly at energy levels, it may exemplify that thermodynamics maintains the relationship between the units of objects through the transference of thermal energies due to the average outcome of constitutionally set working energy elements that privilege in an object/s represents as a mass-energy frame/s.

{(Here, a little more may clarify that the objectually presented physics values expression thermodynamic mean, 'thermo + dynamic' is an emanated overall outcome due to constitutionally representing the energy elements like the temperature- element to generate sensibly determining cold or heat form of elemental values, and the Motional values of 'motion element' generate motion of its' high or low degree of assessment when the functioning object reasonably accepts as a mass-energy

frame, ***qualitatively*** *contain an appropriate form of energy.*

(Here, qualitatively means the representing object or the unit-value energy integrally must occupy such class of energy that qualify to emanate energy-values for thermodynamic work in some conditional atmosphere)}.

i. Herewith are some technical views at energy levels constructional aspects of the 1st-energy-belt of the Universe.
Consequences of innumerable releases of energy - elements, or simple manner may say that the energy values work effectively organize the different sections of the 1st-energy-belt from baby to growing stages of the Universe.

Besides, diverse acts of elemental - values of innumerable energy-elements or overall, as a unit or union of 'mass-state energy' may speak multitudinous acts of energy-values *(or objectually familiarized sense as physics values)* systematic works based on Universe-Time-clock drive towards an ultimate and gradually upon maturity, led the *1st-energy-belt to* reach an irreversible position of the Universe.

→Among the different sections of the 1st-energy-belt, the peripheral territory or the peripheral line, which is the next or alongside the Supreme - Universe, and the containing units of energy in the peripheral zone on the aspects of 'state-of-Absoluteness' property, almost have no differentiability compared to units of Pure-energy.

Also, following an analogous strategy, from the standpoint of energy and physics values application or on the aspect of diversity in the 'baby Universe, the peripheral zone of the preparatory Universe, virtually on 3D aspect reached nearness.

But, practically on the point of energy, the periphery became equal to the 'state-of-Absoluteness' featuring the Supreme - Universe.

Accordingly, on the point of effectivity or may speak, computable energy times for the 'Universe-Time-clock' cannot be applicable for the peripheral domain since it reached a zero-reckoning energy-time *(which is somewhat similar to the conceptually imagined 'truly no-time' state-of-Absoluteness)* along with the other territorial assets of the 1st-energy-belt.

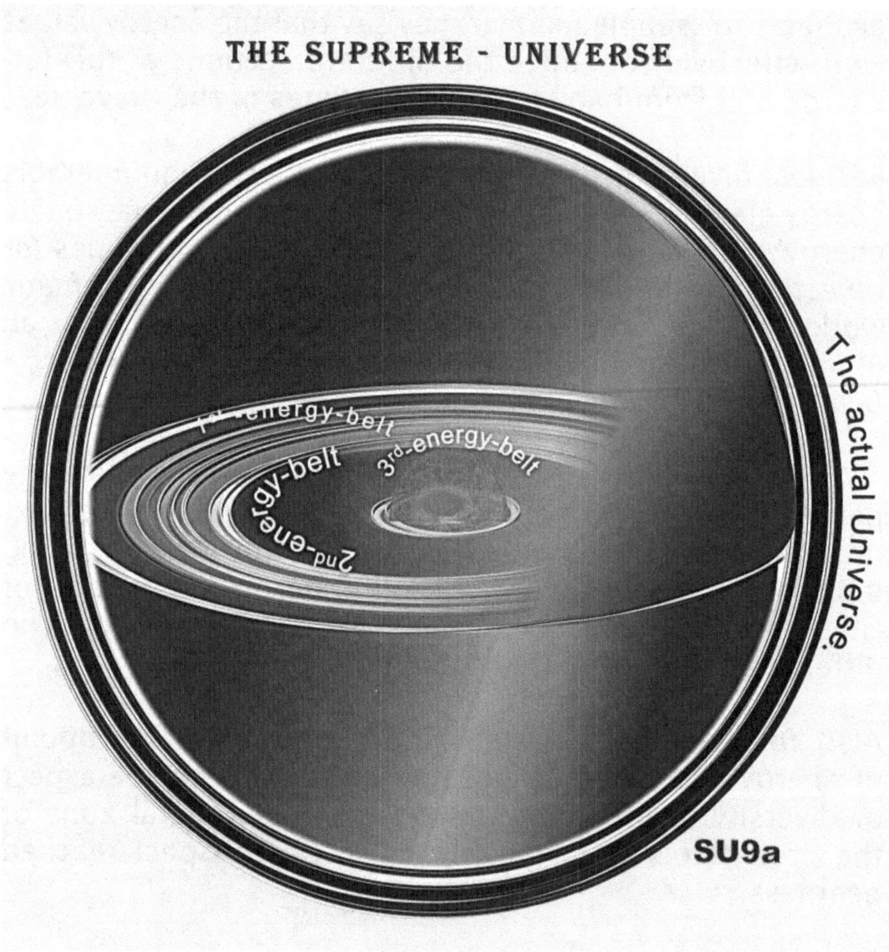

→For the 1st-energy-belt, the two shielding barriers are the 1st-shielding barrier and the 3rd-shielding-barrier endowed with the permanency of the Universe. Besides, it gives the security of the 1st-energy-belt from the perennial turbulent activities of energies in the 2nd-energy-belt.

But in short, it is to say, the distinctiveness on the point of overall quality and quantity of energies (*or the same reasonably justified in an observable sense as elasticity and flexibility*) or characteristically the 2nd-barrier, which seemingly entirely isolated from the rest of the two other barriers that is the 1st-barrier and 3rd-barrier represent as strong shields for the 1st-energy-belt.

→Further to the drawing SU9, the drawing SU9a serves as a theoretical or visual representation of the structure and evolution of the actual Universe. It proposes an unconventional philosophical approach to understanding the actual Universe at its most fundamental level, suggesting that "afflicted or restless Pure energies" energetically guide and organize their position within the Supreme Universe.

Chapter 10

"Energy levels' technical aspects of the Progesterone and the Progression of the Supreme- Universe."

Although the word sense Progesterone *(that is a type of hormone)* is a biological term utilized mainly for female conception, the same in a remote sense applies to clarify as the implicant that allows for initiating the **lymph**/s (*drawing SU10*) for the process and forming the fecund-Universe from the mother Supreme - Universe.

{(By the way, it may be well conscious that the biological word sense progesterone is a type of hormone that performs as a chemical messenger or the same chemical emissary, which is possible to imagine distantly, to imply at energy levels some forms of most early stages' energy having distinctive functionalities to solve some sorts of progesterone purposes that play vital roles in generating offspring of the mother Supreme - Universe.

So, based on the same biological sense, comprehensibly by implicating somewhat analogously may speak that some form of energy elements engendered /stimulated immediately upon initiation of diversity due to the non-sequential 'lowest effect sliding motion' of affected Pure-energies that specifically perform the works of energy levels progesterone in the womb or the prospective IEfu of the mother Supreme - Universe.

(Although, technically, at energy levels, the conception of baby Universe tried to elaborate in Part 1 of this book series that will fall far away when someone attempts to compare following the standard way of matching the medical term 'conception' of living beings.)}

i. For eons of energy times, due to the non-uniform motion of unit-value energies. Or, say as various eventful activities of energy elements, including the motion elements inside the **Influencing-energy** of the "fecund-Universe" that is 'IEfu' derive with appropriate or near-enough incessant splitting and merging process of energies inside the pre-embryonic spatial of the Supreme - Universe.

*(Here, the sense of '**Influencing energy**' indicates that at the primary stage or prior to the commencement of the progesterone, elementary levels of some proprietorially distinct energy elements or 'influencing-energy elements,' besides the perpetually existing motion and motion elements, actively swayed or getting dominated over the 'IEfu' areas in the Supreme - Universe.)*

The eventful activities of affected Pure-energies are the primary cause of inducement and initiation for forming some **unique items**, and due to consistent most low levels elemental values or physics values functionalities consequences, derived that rare or **unique items** introduced with early stages **lymph**, **lymphatic-energy-bubbles**, **lymphatic-clump, and lymphatic energy-mass or lymphatic-mass** formations, along with many supplemented innumerable and frequently appearing **mysterious** energy-elements that is, "**e-el^∞**" times as the acting components for carrying on the process towards development from initiation of the actual Universe.

*(Here, the word-sense '**mysterious**' is applied since the energy levels or more deep-down elementary levels of energy elements works are so instantaneous, it will seemingly or factually be a mystery if attempt to analyze following an observable sense to the standardness of contemporary living species.)*

*(Also, the "**e-el^∞**" times applied to indicate 'innumerable and frequently appearing mysterious energy-elements or energy levels working essences,' based on the delimiting perceptible sense to intelligent species in this biological nature and many other alien natures.*

Besides, technically, one of the most significant findings to remember about the 'law of energy transformation' following the diversifying roles of accessible objects, which factually proves that the appearance and reappearance of "energy-elements or energy levels working essences" are substantial instead of believing or judging it as a mystery, maintaining the objectual and observable pre-set measurable items as constant to reserve the standardness of ideologies of human species and their designed ideology creating machine-like AI.)

1. "Some notes with a drawing SU10, on Lymph-bubble or Lymphatic-energy-bubble, Lymph-clump, Lymphatic-mass and so on in Influencing energy zone (IEsu) of the Supreme - Universe."

*[{(Here, it is to note that since energy levels' multi-dimensional drawings are more complex, thus the pictorial presentations will look like jumbles. Hence, the drawing SU10 and many other drawings in the book series predominantly portray 3D projections, just for convenience to convince human beings of this biological nature, where at some point in energy levels may be found strenuous and illogical to accept and even unrealistic or fantasy through the natural and some unnatural level of computational feeds for measurements that fundamentally acts as preset values (*or, like predeterminable constants*) for overall satisfaction.)*

*{(Also, one point to remember from part 1 of the book series is that the inference of stagnant 'motional-value' is the root cause of gleaning disproportionate and pre-preparative incitement in the prospective **IEsu (IEsu= Influence energy of the Supreme Universe)** of the Supreme - Universe before honoring her **gestation**.*

*Here, for ease of understanding, it is to recount that the prospective infectious areas for the preparation of the Universe, that is, **'IEfu' (IEfu=Influencing - energy of fecund Universe)**, is also beheld at Absoluteness-status in the womb of mother Supreme - Universe.)}*

*{(By the by, the standard term "**gestation**" primarily applies to mammals and refers to the period between conception and birth, during which the embryo develops in the womb. Throughout this period, the child bearer undergoes many changes, internal and external to their physique, and even brings changes in terms of psychological or non-physical activities. In this context, "**gestation**" is specifically used to describe the mammalian reproductive process within this*

habitable environment, which differs from other reproductive or procreation techniques found in non-mammalian living systems.)

{(From a broader perspective, the word "reproduction" can be considered as replicating something, bearing some resemblance to the original.

*So, from the viewpoint of structural configuration like an ascribable object, one could think of replication, identicality, or the production of a positive image or **thing** that is possible due to favorable energy levels aligned with the surrounding atmospheric conditions.*

It also may mean that the internal energy levels physics of the participating object and the external atmospheric physics co-ordinately could manage a replication or reproduction of a bona fide piece of object.)

*(Here, the expression **'things'** is somewhat referable as an object or objectual item representing a form of mass-energy frame/s, so based on the observable sense, it* (replication of thing/s) *does not directly mean 'energy' since the True form of energy is invisible.)}]*

The naturally borne word-sense **lymph** is commonly considered a thin but coagulable fluid.

However, the word '**lymph**' and its resonating expression may find it comfortable to utilize the identity of foremost changing **unique** or **mysterious modules** that release and mislaid during unusual symmetric and asymmetric activity among the break-off units of infected or affected Pure-energies in the pre-embryonic areas for the progression of progesterone in the Supreme - Universe.

But, based on the objectual sense, in the initial stages in the zone of 'IEsu,' features of those **unique** or **mysterious modules** are as good as non-reactive & ineffective semblance as purity of Pure-energies.

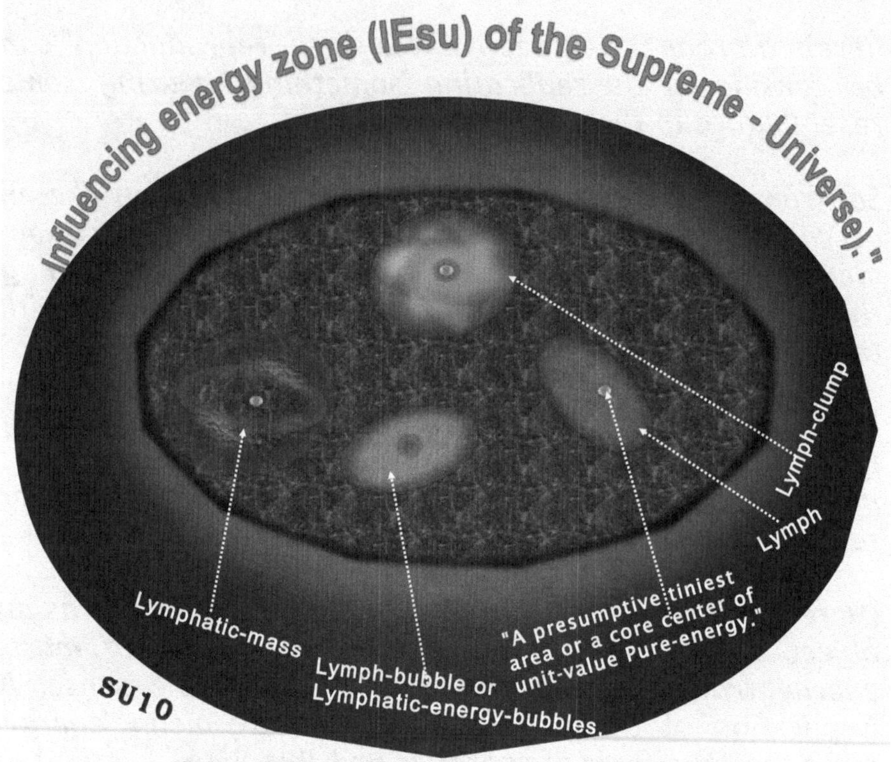

The 'no beginning and no end infinite' 'state-of-Absoluteness' of the Supreme - Universe

→From a technical viewpoint, the **surface tension** and coercion of afflicting-pure-energies on the preparatory **unique** or **rare module** units leverage much more for formation and progression as **lymph** states.

Here, briefly speaking, the primary stages of unusual progesterone activity in the afflicting-pure-energies and the retaining Pure-energies of pre-embryonic areas or 'IEsu' in the Supreme - Universe is the main reason for actuating like **'surface tension'** within the delimited boundary that spread over throughout the 'IEfu' of state-of-Absoluteness featuring Supreme - Universe.

*(By the way, regarding the standardly acknowledging word sense about '**surface tension,**' here may be represented a little differently for convenience to understand energy-level works.*

And briefly say that the 'surface tension' is some form of energy-physics or physics values application concerning the positive (action), negative (opposite), and possibly many other assessable dimensional values of energy level's 'elemental-force' or 'elemental-coercion' that perform some works, under the consistency of conditional energy-atmosphere to the surface of presuming unit-value mass-energies.

Or the same correspondingly may justify as a surface film of an object like a liquid unit representing a unit-value mass-energy frame under some standard or natural atmospheric condition.)

→Also, in an introductory manner, speak that there are celestial, natural, mythological, philosophical, and scientifically expressed standard ways of expressions by using words, besides possibly many other fuzzy senses and discernment, accessibly and reasonably utilized for application and to explain energy levels functions or works.

Accordingly, from the viewpoint of energy, many standardized and unstandardized expressions that apply in words, gesturing, and many other physical and non-physical techniques for natural and celestial purposes, which may be reasonable to employ for ease of comprehension in specific ways positioned intelligent genera like contemporary intelligent bio-beings in this biological nature and possibly other natural ways adapted isolated or alien worlds of different 'cosmical-Universes' in the matured 2nd-energy-belt.

Also, in some cases, those natural ways and unnaturally developed expression techniques are partly or utterly inappropriate and impractical to comprehend by the non-local or distant living intelligent genera or species following the technicality of physical constitutions and dissimilar practices in a suitable life-orienting survival atmosphere.

[{("Some analytical clarification."
Whatever, from the viewpoint of energy, the mean of all those set forth words and senses and their purpose of expression have significantly direct or remotely linked throughout the Universe, either regarded as conceivable or inconceivable for the case of the representing tiniest or the gigantic form of living and non-living products /objects, following the individual or group constituent.

Technically, the above may be introduced as a relativistic relation or philosophically may present as relativism since energy is everywhere either this or that form on the point of particularity for discernment of specific way designed human beings and alien intelligent species.

*Accordingly, it will be logical cognition that the objectual word sense or expression **'causal connection'** is appropriate to apply for making knowable and unknowable energy levels links throughout the matured Universe, mainly the extreme diversifying works in the matured 2nd-energy-belt.*

*{(For technical clarification related to the point of **'causal connection,'** it proves the authenticity of the famous equation E = mc^2, where 'E'* (that is, 'Energy'*) maintained unchanging relations for all sorts of diversity served by the 'm' and 'c' within some specific areas indicated by ^2.)}]*

i. In brief, technically, consequences of eventful motion due to progesterone activities in the 'IEfu' derived devaluation and gradually **weakened** the **surface tension** following elemental loss from the afflicting-Pure-energies.

Simultaneously, the growing lymph begins to encounter internal resistivity due to the elemental roles or elemental physics derived in the containing or ingrained energy elements represented in lymph forms, that is, some states or units of 'mass-energy' which usually deters cohesion smoothly with other Lymphs.

{By the way, corresponding to the above, in an objectual sense like liquid, it remembers one of the standard points of ***surface tension****, which (*that is, **surface tension***) gets* ***weaker*** *upon increasing physics values like temperature.}*

→However, some ineffective unique items and units of afflicting energies around the Lymphs, externally acted as **assistive - coercion** for cohesion to change the original structure of lymph to turn into a **lymphatic-energy bubble**.

(Here, ***assistive - coercion*** *is employed to indicate additional physics values like coercion from an earlier assumptive point or value in the progressing 'IEfu').*

→Further, primary stages Lymphs are characteristically carefree and comprehensively denser from subsequently liberated unique items or lymph-forming energy.

→Technically, the release of unique modules in the vastness of energy means derivement for initiation and ill-treatment to the originalness of other unaffected Pure-energies that held and rested in the 'IEfu' or prospective baby or fecund Universe. It also means that the increasing sequential diversity consequentially upholds some form of effects to almost all units of Pure-energies that are clung in the IEsu.

→Although at energy levels' everything works multi-directionally.

However, analogously *(based on object and energy),* applying the common observable mean or sense of **'non-reactive** and **non-effective**' may speak that this technical sense of **'non-reactive** and **non-effective'** characteristics became ease and apt to understand where there is any internal or external logic inference from a firm objective and the contending unit/s of energy, which may nullify, impair, or dissolve if any creation due to deferrable energy-values or in-depth of elemental - values in a transitory, momentary or time-bound manner to the standard consciousness of intelligent human species.

→But, for the primary stage's units of **Lymph energies**, the objectually applying features like non-reactive and non-effective are not wholly appropriate to imply since there are not enough reasonable contenders like the observable sense standardly acknowledges.

The same, simple outlying in a socio-scientific ethical manner may exemplify that those babyhood **Lymph energies** are under the **gravity** and guardianship of forebears and so encounter insufficient contenders or comprehensive differentiable - times that may project with objectually presented sense 'non-reactive & non-effective.'

(Indeed, objectual definitions are seemingly isolated and different to catch through the standardness of living beings, but the 'physics' is physics, which eternally maintains an equivalent or interlinking relationship with everything on the aspects of diversified forms of energy initiated by the birth of the Universe to its contemporary eternal mature status that defined by the unified equation, or speaks as a whole, the most compact state formula or universally unified formula 'E = mc^2.'

So, from the standpoint of energy, the Great-grandmother Supreme - Universe, the Universe, and their all sorts of progenies retained imperishable relations whatsoever intelligent species negate to acknowledge through the delimitation of observable or logical practicalities.)

→Therefore, following the above explanations, it may characterize that the formation of Lymphs is one of the root causes for deriving enhancement at the preparatory phase's commotions among the retained unique items, limps, hobbles, and afflicting-pure-energies in the IEfu since increasing the disproportionation of **physics** compared to the many other neonate energy products from the afflicting-pure-energies and the infinite coverage of 'IEfu' in the Supreme – Universe.

►Gradually towards progression and upon getting the individual status as Lymphs, seceding-units-of-energy that is the 2nd-stages afflicting-pure-energy, remnant unique-module, limps, hobbles and some relatively indolent supplemented-Items inhabits /occupies much differentiable status.

→Whatsoever, technically, following the standardness of intellectual conceptions, the Lymphs are seemingly more active but less effective than the enormous seceding units of afflicting-pure-energies in the vast progesterone areas.

{To ease comprehension of the above points, a drawing SU11 may be projected to indicate the earliest phases of energy physics or may justify as in-depth of most momentarily deriving elemental physics of energy elements that endowed towards total development to everything in the IEsu that gave the life of the baby Universe, which may summarize and partly introduce with the following energy levels technicalities: -

a. *Analytically, the Supreme - Universe is a state-of-Absoluteness, a state of the true singularity of Pure-*

*energies, so the 'question of physics at this stage may symbolically present at its supreme state as **abs0^∞'** since the Supreme - Universe is a no beginning and no end status. Thus, the objectually recognizing 0 /zero and its absolute (abs) position comparably will look like a kind of observable satisfaction only.*

(Also, it may give a little clarification and speak that intelligent living species like human beings mostly tend to discover or compute the limitation of direct and indirectly assessable items, maintaining oneself and the agreeable group or species as a base, constant, or 0-like symbol for analyzation and satisfaction.)

b*. In the Supreme - Universe, fundamentally 'physics' initiated upon the disruption of regular symmetric null-effect-sliding-motion and the appearance of the unique state of rare appearances Asymmetric nonsequential null-effect sliding-motion* (this point illustrated in part 1 of the book series).

c*. The more the affliction or disorderliness of the pure status of Pure-energies in the IEsu, the increasing the diversity due to most momentarily liberation or the appearance of elemental values* (or the physics values*) of energy elements that integrally evolved as the singularity of the Pure-energies.*

So, the increasing of 'physics' resulted in enhancing the effects with the formations of unique items or energy products like Lymphs, limps, hobbles, and so on (from the most momentary to long exhibiting) *to infinite in the vast ocean of energies of the IEfu.*

► *Besides, for the convenience of understanding, one thing may touch and speak that the physics values roles gradually aggravated /intensified by* **diluting** *(or, this objectually applied word-sense* **diluting** *may justify as reducing or may be considered as seceding the* *integrating strength of Pure status of unit-value Pure-energies) the elemental values of energy elements from the original status of Pure-energies within the limits of IEsu zone, which is entirely unlike the observably acknowledging sense as density that usually measures to some logic limits of the parametric degree to determine the word-sense expansion and contraction, explaining something like 'vacuum or nothing' that accomplishes as constant to all around the unfathomable or infinite areas coverage of energy.}*

'The Supreme – Universe, and objectually justifiable if any physics like the supreme-physics or 'abs0^∞' in the no beginning and no end areas Pure-energies.'

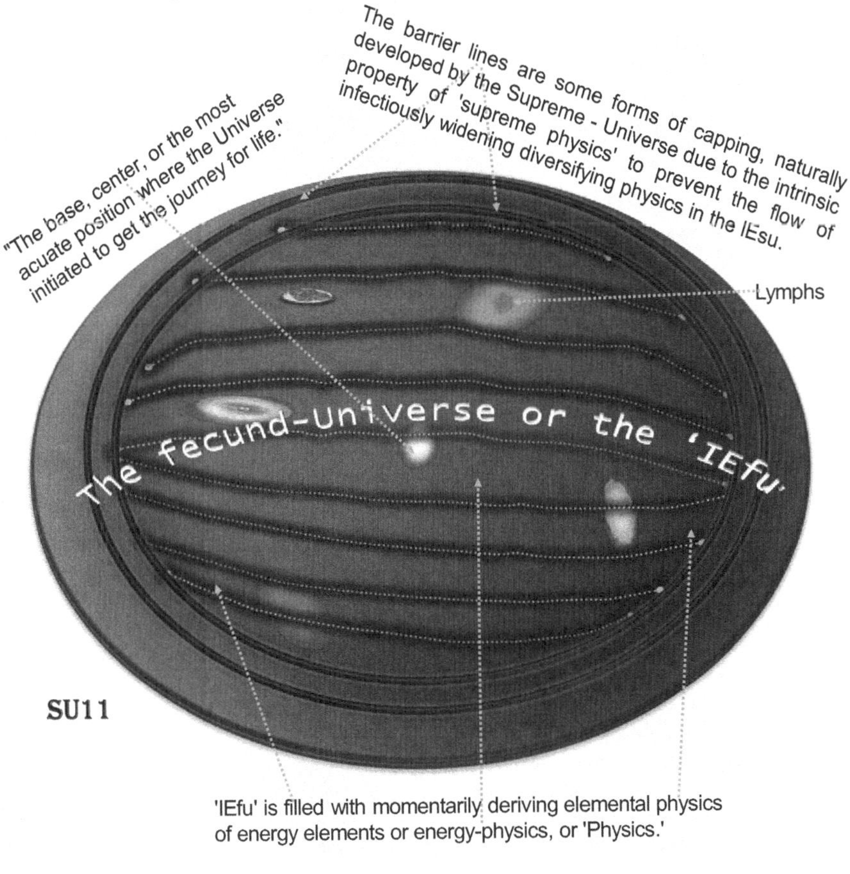

SU11

→In drawing SU11, the following points are marked, besides adding some analytical clarifications on the related issues.

► 'The Supreme - Universe, and objectually justifiable if any physics like the supreme-physics or 'abs0^∞' in the no beginning and no end areas Pure-energies.'

*(Indeed, from the standpoint of energy, the Supreme - Universe is a 'state-of-Absoluteness,' so all the ideologically presentable physics and consequential diversity can be justified as 'quite compactly hinged into a fused unit like **abs0^∞**' in the form of godly honor Pure-energies.)*

►" The base, center, or the most acuate position where the Universe initiated to get the journey for life."

(The indicated base point or the most acuate position, considered an origin from where diversity commenced from a very early stage's progesterone-like function that gradually progressed with increasing physics values roles.)

► The barrier lines are some forms of capping, naturally developed by the Supreme - Universe due to the intrinsic property of 'supreme physics' to prevent the flow of infectiously widening diversifying physics in the IEfu.

*{(The barrier lines unpretentiously represent superiorly formed prophylactic areas due to debilitated internal physics values functionalities of IEfu and the most substantial impediment of surrounding **'supreme-physics** or abs0^∞' of the Supreme - Universe.*

*Technically, **supreme-physics** or abs0^∞ of the 'no beginning and no end' may be simplified following the delimitation of observable sense to intellect species in this bio-globe and say as a multi-dimensionally representable True form of 0-status physics of Pure-energies that allowed*

Supreme - Universe to prevent the flow of infectiously broadening physics of the IEfu.

(Here, one thing to remember for clarity of observable comprehension that standardly justifies 0 /zero is a consolidated ceasing status, where computational interpretation follows some or delimited befitting dimensions, maintaining the vacuum, empty or nothing like sense for a fictitious constant that may satisfy an individual or their group as a species for some reasonable findings, which is entirely unlike the momentarily changing dimension for energy levels works in the 2nd-energy-belt.)}

►'IEfu' is filled with momentarily deriving elemental physics of energy elements or energy-physics, or 'Physics.'

[{("Some analytical clarification."
By the way, for clarity, it recalls by quoting that 'energy-physics' or in-depth of integrated most momentarily deriving elemental-physics, infectiously keep on ***diluting*** *in the IEsu zone.*

Or objectually, the same sense ***diluting*** *may present as* ***'widen*** *and widen in the vacuum or nothing like pretending constant' consequences with many outcomes that drove toward some ranges, before checked due to the self-potency of state-of-Absoluteness or abs0^n of the Supreme - Universe.)*

{It may requote for the avoidance of any confusion that energies are everywhere, so factually on the point of 'energy,' dilation and constriction (that is, observable **'widen** and shorten') *are just like parametric variations from some lower and upper degree of measurements, maintaining two parks as some level of thinner /lighter assuming to some lower and the thicker /dense assuming some upper, on the point of scaling.*

▶ *Indeed, technically, the above clarification is possible to wrap by the equation E = mc^2,*

where 'E' can be a judicable constant that maintains an equivalent relation of coverage indicated by the ^2, containing mass (m) of some higher degree than the (c) comparable lower degree of energy in the unit value, indicated by the symbol 'E' the representing constant as a whole area or coverage (that is, ^2) occupied by the 'm' and 'c.'}]

ii. By the way, in a reasonable sense, it may present some simple notes concerning the word sense, Gravity. The drawing SU12 is just to refresh the standard conception of Gravity. There, it indicates the well-known findings about an apple falling from a tree due to the objectually recognizing touchless coercion of earth.

(By the way, it's worth mentioning that some significant issues related to "Gravity" have already been elaborated in the book Energy and the Complete Universe, Part 4. It includes a detailed explanation of how Gravity functions within the context of football association.)

→Indeed, on the point of energy, 'Gravity' is a constitutional part of this bio-globe, so its pulling coercion mainly applies to its native everything regardless of living and non-living micro to massive structural *(observable and non-observable)* articulations.

→Presenting a comprehensible example (*shown in drawing SU12*), where the bird can freely maneuver, but the apple is falling from the tree in a reasonable atmosphere of surroundings when some degree of earth's gravitational action is persistent on each of them.

→Here, for a short clarification, it assumes that the apple and the bird occupy the same amount of 'mass-energy' referable to individual identities as an apple and a bird.

►Technically, from the viewpoint of energy, immediately upon release from the pedicle of the apple, the apple begins to fall, which was held by the branch of the tree.

a. Analytically, the apple clings with the branch of an apple tree due to the functioning of one of the fundamental controlling parts of the tree's physics that allows for the maintenance of the apple's overall *(means internal and external)* physics in the surrounding conditional /standard atmosphere.

Immediately upon losing one of the naturally provided fundamental physics to retain the apple in balance with the liveliness activities of the tree, the apple **healthily** will begin to fall for Earth's gravity.

{Regarding ***the healthy condition of the apple*** *is due to the apple's intactness of the rest* (here, rest means isolated status from the tree's physics*) of the constitutional mass form energies and physics as a referable frame that gave the identity as a fruit.}*

b. But, for the bird, his structural value and survival habits developed in abidance of the natural laws of this bio-globe, so the constitutional physics grown by birth is well set to maneuver under its (*bird's*) healthy condition in reasonable surroundings.

Yet, the earth's Gravity acts equal to the apple and bird since both are assumed to have the same amount of 'mass energy,' but considering the structural aspects of referential frames like an apple and the bird differ from others, which allows them to follow the standardness of individuality and capability of maneuvering in the conducive surroundings.

c. Indeed, while the child sentiment views the falling leaf with a sense of quiet melancholy, from a more resonant philosophical perspective, the leaf embodies a duality of existence. On the one hand, it may take pride in its individuality, striving to establish its self-laws for survival, ever adapting to the squads of nature that shape and transform its being. Yet, on the other hand, it may grieve its separation from the parent tree, longing for the nurturing connection and shelter once provided, now lost to the winds of change.

►Analytically, some technical clarity on the aspects of energy levels Gravity.

In a cursory manner, it may say, in this bio-nature, the word sense gravity, in general, is defined as coercion /force that tries to pull two objects toward each other.

Also, following living ethics, the same word Gravity may mean the positive and negative sense of allurement or attraction and repulsion or obstruction features that satisfy specific ways evolved intelligent species.

→Whatever the logic comprehension following the trend of observable accessibilities based on this biologically evolved species, justify either positive or negative ways of discovering or remaining as obscured on the sense of Gravity.

From the standpoint of energy in the matured 2nd-energy-belt, Gravity is a certain kind of energy level essences, which posteriorly /latterly get acknowledgeable entities in objects and objectual products, and like the advanced phases' matter-energies, matter-elements, matter-products, and so on new-new products and byproducts or overall may speak them as advanced stages progenies that represent for unit form mass-energy frame/s, maintained relative relations from most-micro to the massive distinguishing status.

→So, it is reasonable to say that any referable mass-energy frame/s (*that is, the object*) accomplishes some form of individual roles based on the functional outcome of its constituent (*integrated*) energy elements, energy components, or the most momentarily functioning working essences, which can simplify with one of numerous **proprietorially** distinctive energy elements like the typicality of Gravity generating physics.

→The above earthly ways of discernment are entirely unlike or isolated to justify the utmost primary stages of energy elements in the ever existence boundless Supreme - Universe, where all energy elements or working essences stay as one and only one for attribution to give godly honor of **Pure-energies**.

Or, for computational convenience, this Pure-status of Pure-energy may be remotely analog like the objectually implied physical or objectual understanding as absolutely 0, which standardly represents a conditional absolute zero-value or position, functionally acts as a null for the outcome of conceptional sequences for the intelligent system, acclimated in this bio-Globe.

(The above-mentioned objectual word sense ***proprietorial*** *means 'the fact of owning something' to an individual or their group as a species.*

*The word '****proprietorial****' is utilized only for convenience and to ease comprehension of energy level behaviors of momentarily changing energy elements, along with their ownership or proprietorial roles in any form of considerable mass-energy, mass-energy frame, or object.*

Besides, one thing may refresh from the standardness of conception that the elementary level or energy levels' physics will seemingly be entirely unlike while assessing and analyzing the daily accessible objectual physics that performs through some naturally evolved mechanism, including living beings and non-naturally developed measuring tools within the limits of intellectual perception /horizon.)

→Following the standpoint of energy in the 2nd-energy-belt, a little more may clarify for the convenience of understanding that all forms of unit-value energy, mass-energy, and mass-energy frames at their early and advanced stages, regardless of representing energy products like objects or microparticles may or may not be capable of delivering a reasonable degree of the gravitational outcome, which go with or adopt on the quality of constituent gravity generating components, like energy-graviton and the surrounding conditional atmosphere, that is, energy-Atmosphere and if any Twig-atmosphere.

→So, it recites in short and speaks that the quality of **unit-value** energy (*indeed, the objectually presented word sense* ***'unit-value'*** *indicates some reasonable, lowest or highest level of quantity*) referable as a mass-energy frame or an object, and their integrally positioned energy elements or working essences define the **yield and behavior** for performance in concordance with the surrounding atmospheric circumstances.

{(*Here, the objectually conveying consolidated models of the word sense* ***'yield and behavior'*** *seemingly to be quite unlike when attempts to analyze directly through the in-depth energy-levels physics values roles of objects, following the standardness of classically developed intellectual conceptions and computational practices.*

But, from the standpoint of energy, all sorts of physics causally derive the outcome as ***yield and behavior*** *somehow* (means, directly or indirectly) *reserved interlinking relation throughout the energies in the 2nd-energy-belt.*

Therefore, the parametrically estimating word sense 'extra' dominates its importance or values for objectual applications only since the objectually determinable word sense quantitative is wholly settled within the 2nd-energy-belt, gradually and immediately reaching its mature state in the actual Universe.)}

For the earlier mentioned topic of **'yield and behavior**,' it may be reasonable to exemplify by remembering the standardly or classically acknowledging differing strengths of Gravity to different planets (*shown in the drawing SU13*) based on the aspects of some volumetric or dimensional conclusion that follow the standard objectual computations.

However, at the energy levels' Gravity follows the in-depth of integral energy elements or elemental composition and **quality** of the amount of mass-energy referable to individual planets in the vast energy -Ocean of the 2nd-energy-belt, which *(that is, quantity and **quality**)* applies to everything regardless of proprietorially distinctive mass-form energy, representing their unique identities as most micro to massive structural recognition through the standard conceptions of living species.

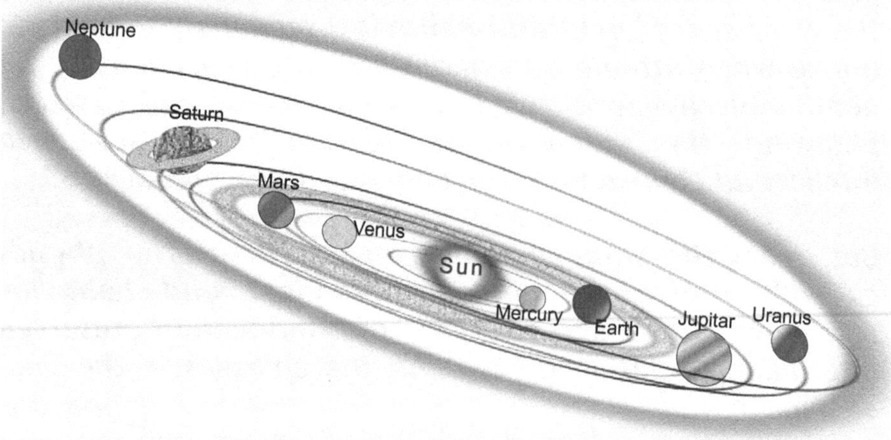

SU13

→Therefore, if not any existence of energy-graviton as an integral part of unit-value energy, mass-energy, and mass-energy frames at their early and advanced stages, which will represent energy products like small or large objects, there will be no functional effect on the point of gravity-related role, like the 'apple and bird' as mentioned above in the drawing SU12.

Now, for the apple, the representing frame is a sack of mass-energy compositionally integrated with some appropriate types of energy elements, including graviton, so it is capable of maintaining the healthiness of its molecular union but inept of flying under the standardness of the surrounding atmosphere.

Whereas, even assuming that the bird occupies the same amount of 'mass-energy' as the apple but integrally set up with proprietorially distinctive energy- elements *(besides the graviton)* that allow the bird to check /prevent from free fall and reasonably maneuver when flying, maintaining internal physics of physique under the same atmosphere of apple's free fall.

{(Indeed, *it is agreeable to every intelligent being that mother-nature bio-globe successfully could train to develop like the bird, ant, tree, apple, and likewise many, in this life-orienting habitable zone for survival requirements following natural laws of physics that applicable in-depth of elementary levels of any unit-value energy in a place of appearance, maintaining amenable relation of universal physics that cover throughout the 2nd-energy-belt.)*

(Besides, applying the in-depth energy levels senses, one thing may say that the tree, leaves, and all other surrounding visible and invisible objectual products are causally related to the reasonability or conduciveness of internally generated physics and the external physics of surroundings. Wherever the products shelter in a natural atmosphere or tend to transit through a far distance of foreign atmosphere in the 2nd-energy-belt.)}

[{("Some other energy levels' technical clarity about the appearance and evolution of unit-value energy, mass-energy, mass-energy frame, or object."

For ease of comprehension of the above, a little deep-down of the energy levels of the matured 2nd-energy-belt.

(It attempts to indicate with a drawing SU14 that if it considered making any unit-value energy at its most micro

possible state, there must be compositionally two minimum forms 'energy elements.' Like the most fundamental motion element for deriving motion, the other one will be the influencing energy element for generating leverage-like essences, so there will be no chance of deriving gravity since there is no third-one 'energy element' or working essence like graviton or pulling like roles in the unit-value energy.)

{(The SU14 is sketched with two outlines to indicate a possible simple and some state of a complex arrangement of energy elements in a unit-value, a packet, sack, or mass state energy.

►*One is a simple depiction with four mono-state elemental positions of energy-energy elements, centrally located 'motion element,' and three layers occupied by three proprietorially individual energy elements.*

►*The 2nd one is, comparably, an aerial and **transparent** view of a complex arrangement of nine layers of stepwise increasing sets of 'energy elements' in a unit-value energy.*

→*Analytically, in the 2nd-drawing of SU14, '**transparency / translucency**' indicates some levels of energy covered throughout the unit value, which proprietorially act similar to the feature of 'ostensive Purest-energy' or the same may mean some state of highest dynamism and neutral roles like the constitutionally set motion element* (centrally located*) of the unit-value energy, mass-energy frame or an object.*

Here, just for conceptional clearance, it is to recollect that energy is unlike the features from classically presented various states of matter like solid, liquid, gas, and plasma, which are easily determinable following some parametric sequences.

Technically, any imaginable amount of energy undergoes lots of most-momentarily changing physics-values functionalities before it transduces into a peripherally distinct unit value for representation as a mass-energy, mass-energy frame, or object. So, classically speaking, the sense of 'flexibility' at energy levels is utterly incomprehensible compared to the parametrically preset or delimiting mindset of modern intelligent beings.)}

(Also, from the perspective of energy, I would like to mention in brief that when someone thinks of any unit-value energy at its most minimal status considering some form or level of stableness, there needs to be one more with the foundational two 'energy elements' those are the 'motion-element and influencing energy element' since energies in the matured 2nd-energy-belt are intensely diverse, reserving the constancy of fastest and neutral roles of ostensive Purest-energies.)

(Besides, it is to speak that the ostensive Purest-energies proprietorially represent as mono featuring motion on the aspect of 'peakest dynamism' and 'neutral' on the point of effect in the matured 2nd-energy-belt. So, the ostensive 'Purest-energy' is considered a 'perpetual constant' instead of standardly acknowledging multiple types of differential constancies for measurements of all objectual and non-objectual assignments that are accessible through the systems of living beings.)

Above and beyond the endurance of the fastest and neutral roles of 'ostensive Purest-energies,' it proprietorially functions as an inexhaustible and the most fundamental regulator for everything within the periphery of the matured 2nd-energy-belt, surrounded by the 1st-energy-belt and the 3rd-energy-belt since the ostensive Purest-energies represent as a universal base, which is unlike the objectually applied sense as a temporal base or constant in the vastness energy-Ocean.)

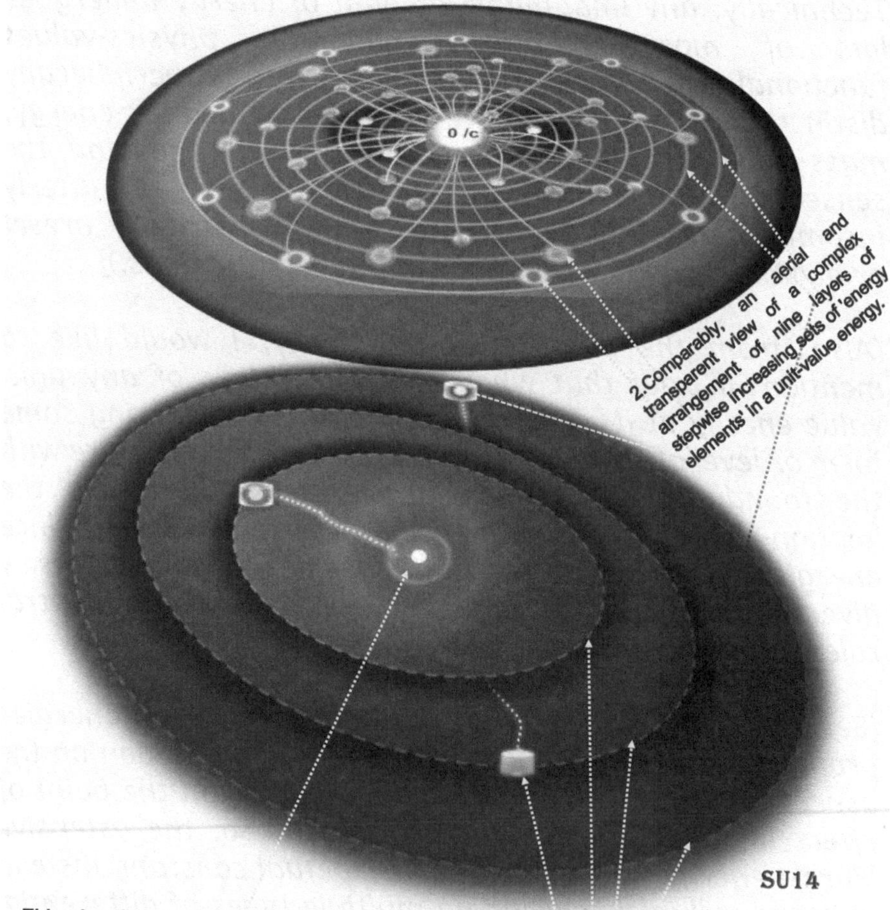

This drawing is a simple depiction with four mono-state elemental positions of energy-energy elements, centrally located 'motion element,' and three layers occupied by three proprietorially individual energy elements.

(Further, say that atmospheric physics alone cannot act perfectly to make stable unit-value energy from the flimsier mass form energies, comparing the sturdy large amount of mass-energy, mass-energy frame, or object.

So, conduciveness or under certain suitable circumstances of external atmospheric physics always looks for top choices for convenience of shaping and organizing the energy elements in a prospective reasonable amount of energy to make a unit or their mass-form energy before constructing it to some state of stableness for representing unit/s of the mass-energy frame and then have a better chance to follow for advancement as a healthy object.)}]

▶ Further, to cite that, fundamentally for intelligent species in a biological, and even if for the non-biological alien world, the equation 'E = mc^2' will act as a covert and compactly formulaic presentation, which hiddenly demonstrates that energy 'E' maintained its equivalent (=) relations with mass (m) whatsoever compositionally constitutes with different states /conditions of energy-elements besides motion deriving motion-element (c), which all symbolically delimited by the ^2 that represent like an area. And that area (^2) may cover some discrete tiny to massive unimaginable infinite space distinguishable or indistinguishable through the delimitation of dimensionally determinable perceptions of intelligent living species.

So, from the above, it is reasonable to demonstrate that if 'E' engrossingly represents the containing of the 2nd-energy-belt, then it (E) will maintain an equivalent (=) relation with the ostensive Purest-energy that acts as motion /speed (c), rest energies will be the mass (m) within the peripheral boundary (^2).

{(Indeed, in the matured 2nd-energy-belt, if someone speculates to pile a tiny amount as a unit-value energy, the containing of the unit-value energy will be the mix of multiple working essences or energy elements like a random choice that can perform lots of sensible physics based on the surrounding circumstances.

But, if one can imagine collecting only ostensive 'Purest-energy,' analogously, it will find only one featuring energy element, that is, motion deriving motion element, where gravity and lots of other elemental physics get no direct entity since ostensive Purest-energy occupies extremity of dynamism and neutrality in the 2nd-energy-belt.

Technically, whenever justified to make a ***unit value*** *from a random choice of energies of energy-Ocean, logically, that* ***'unit value'*** *will be determined by some peripheral conditionality regardless of its constitutional compositions in a reasonably small or oversized covering atmosphere.*

So, coinciding the measuring technique of specific way evolved intellectual beings, is to speak based on the perspective of 'elemental physics or energy-physics,' dimensionally distinctiveness unit value mass-energy whether it represents as a long stable frame or not, confronts two phases of dissimilarities in many aspects related to internal (the unit own-self) and external (energy atmosphere) diversities for the maintenance of the peripheral boundary of the unit.

But, technically, from the viewpoint of energy, physics performs multi-dimensionally at the elementary levels of energy elements for making any prospective unit-value mass form energy following the ***conduciveness*** *of surrounding atmospheric physics at a birthplace in the extensive energy -Ocean of the 2nd-energy-belt.*

Therefore, the objectual word sense 'illusive' mostly dominates during the attainable measurement practices.)}

iii. 'Some clarification about the frequently used terms objects and objectual products.'

In an introductory manner, here may present a few notes on the word sense object and the objectual products, which seemingly may differ from the classically stated standard types of matter like solid, liquid, gaseous, and plasma, when discernment and judgment performed following delimitation of some dimensional assessments.

Correspondingly, those earliest and advanced stages' objects, and then most lately appeared as different states of matter that naturally recognized in the form of solid, liquid, gaseous, and plasma, evidently be far isolated when attempting to judge energy level activities following the limitation of discernible directions or dimensions that participate as one of the significant measuring units /parameters to specific way evolved intelligent living species.

Fundamentally, for convenience to comprehend the diversified states of various matter, materials, and likewise many advanced stages subproducts or by-products, correspondingly may facilitate with a common word like 'object,' which apply as a comprehensive justification of all as developed stages micro to massive structurally positioned mass-energy frame/s gives the shape or individuality of accessible and inaccessible objects, objectual-products, matter, materials, and etc^∞ in the matured 2nd-energy-belt.

2. "Now going to cite some technical points about 'helicity-energies,' which employed on the aspects of growing the zone 'IEsu,' with a comprehensible drawing SU15."

Initially, for convenience, It may recite that any unit-value mass-energy integrally composed with types of energy elements, whether term as early stage's mass form energy, before turning into the mass-energy frames that referable as raw or unprocessed form of advanced stages objects, matter or materials, which all, micro momentarily has a trend to 'releases elemental values' or overarchingly consider as 'energy values' (or, this 'energy value may consider as energy physics or working essence for any representing mass-energy frame or object) due to uncompromising elementary levels' functions of integrated energy elements in any form of early and advanced stages unit-value mass-energy, from the birth to the long-lasting existence of the matured Universe.

The drawings assume unit-value 'mass-energy' stepwise occupies one, two, three, and possibly more elemental positions following the most fundamental or common one is the motion element depicted as 'c or 0.' The energy elements can discern following their proprietorial roles, or conveniently, same may distinguish degree-wise following the motion element as a 'base, 0 or c,' like the one that supposes an 'influencing energy-element' as thus grade-wise the influencing energy element can be presumed as '1,' which is the closest to the motion element or 0.

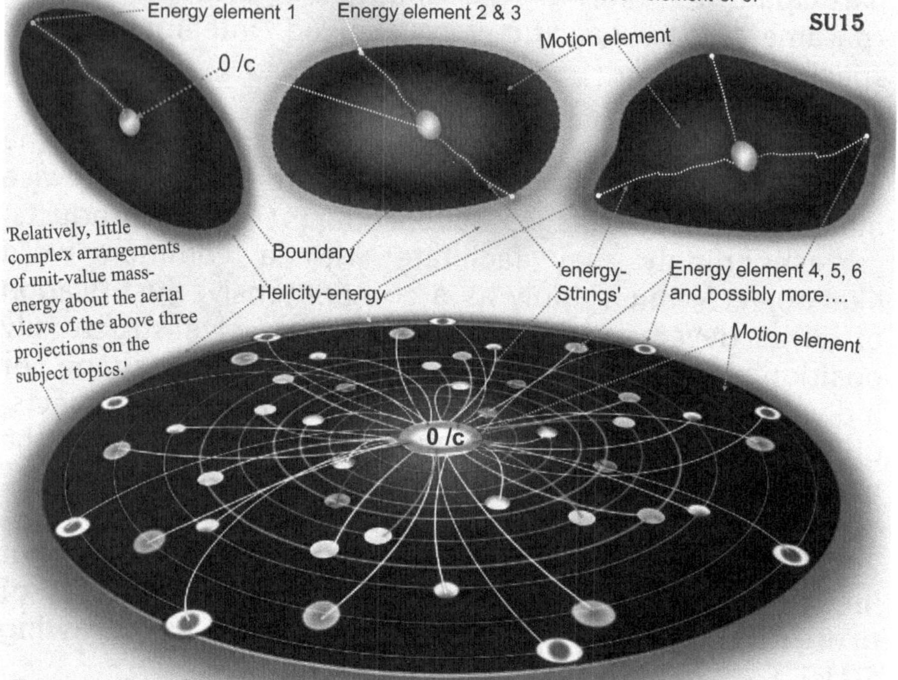

The drawing SU15 with four sets of aerial views indicates the overall elemental outcome as 'elemental values or physics values creations that tend to wreath around as an elegance of 'helicity-energy,' which somewhat performs as a boundary of the unit value when the constituent energy elements engrossingly regulate the internal physics of the unit-value mass-energy.

*{(Here, straightly, the sense '**types**' means proprietorial assignment on the aspects of differentiable functions.*

*→Otherwise, when the intellectual notion is set or rationed to a fixed 0-values position for all purposes, the word-sense '**type**' and many similar expressions cannot stand with long-validated meaning /purposes since the momentary appearance outcome at energy levels tend to represent illusive by nature beyond the dimensional determinism limits of living beings in this bio-Globe.*

Or, the same may present that diversity from any non-fixed or outside of 0-positioning notion facilitates all sorts of differentiable computations and analyses, which ***may or may not*** *perform as fact /valid to some level of satisfaction to specific way designed biologically positioned living beings in this habitable nature.*

*(Here, the objectual word senses '****may or may not****' indicates some parametric presentation following individual personalities, judgments, and superior limits that differ from one another of living beings.)*

As an example of the above, a simple and convincible way may present with the traditionally following objectual words like nothing, vacuum, and likewise many are evolved as pre-set values due to standard practices, based on the structural value as a living being that delimited by sensing faculties to an individual or their group as species.

So, technically, those sensing faculties control all assessable items within some parametric boundary justified by words like 'logic' or a constant that acts as '0-positioning reasonability for all grantable inducements.

Accordingly, beyond the above-mentioned denoted '0' or the parametrically delimiting logic that overall acts as a functional constant for self-measurements for acceptance of an individual living system or their overarching value as a group, characterized as nothing or vacuum-like delusion.}

i. So, for the 2nd-energy-belt, at some specific energy - times, the **energies - released** *(due to constituent elemental values functionalities*) from any mass-energy and mass-energy frame externally **tend** to accumulate to represent the property /identity of **'helicity-energies**.'

(*Following the energy levels sense, the expression* '**tend**' *conveys the possible implication of physics values by obeying one or more conditionalities that differ from the standardly applied objectual word,* '**tend**,' *which mainly* *signifies the behavior or characteristic exhibited regularly or frequently in particular ways.)*

→Further, it may speak, the appearance of those advanced stages of '**energy releases'** gradually, and in consequence of diverse atmospheric physics values reasoning, proclaim as **objectual-energies** for any mass-energy frame or object and their possible discrete units.

Those ready-made **objectual-energies** can be considered some class of advanced stages energies that **smoothly** can transform into new objectual products following the appropriate conditionalities of atmospheric physics, reasonably *(here, the sense reasonably is not entirely or utterly equal)* may match for sharing by other objects.

*(Here, the objectual sense '**smoothly**' practically follows some parametric comparison.)*

→So, it may be the reasoning for extrapolation /projection towards positive, negative, cipher, and other dimensional ways of momentary divergence that introduce and follows the structural configuration and alterations, in concordance with internally generated energy levels physics of prospective unit-value mass-energy, mass-energy frame or object and the surrounding **atmospheric-Physics**.

*(Here, '**atmospheric-Physics**' is justified an overall energy-Physics and any Twig-Physics in a reasonable stipulative area, analog to the vast ocean of energies in the 2nd-energy-belt.)*

→ Therefore, from the above short technical clarifications, it may suit to grasp or realize following the standard conception, devised based on the specificity of constituent compositions of intelligent living species concerning energy levels' continuance physics to express in confidence that anything momentarily can bring the structural alternation. Which technically initiates from deep-down (*that is, elementary levels of energy elements*) of any unit-value energy referable as mass-energy frame or object in abidance and continuance with the rhythm of '**natural-Physics** and **universal-Physics'** of the vast energy-Ocean in the 2nd-energy-belt.

(The technicality of 'natural physics' and 'universal physics' is clarified, as and when appropriate, in this book series and the series ***'Energy and the Complete Universe.'****)*

→ {(For ease of comprehension, here may give a little touch and say, technically, on the energy point of the 2nd-energy-belt, comparably on the dimensional aspects and measurement of functions, any discrete form of unit-value energy or item will fall short or go up than the others in the vast or infinite areas of energy-Ocean, which determine the technicality of expression and authentication of ***'transformation laws'*** *while in consistency with the deep-down of energy levels functionalities.*

*Besides, it (*means **'transformation laws'***) will negate the parametrically determinable observable reality like the sense* ***'equal (=).'***

*Instead, it (*the objectual sense '**equal'***) is appropriate to imply as an equivalent* ***(Ξ),*** *which means that energies maintained equivalent or interchangeable relation with everything in this 2nd-energy-belt whether realizable or not, directly through the own system or indirectly by other mean following the standard perception of living beings, whatsoever easygoing or* ***indeterminable*** *to some micro levels consistent loss or gain are transpiring at the in the depth of energy.}*

ii. Further, to speak natively at energy levels, the word **helicity** is almost similar to means by natural-born particle physics, that is, 'the projection of the spin onto **some** (*here, '**some**' means the limits of objectual logic discretions*) directions of momentum, or assumed as the time-based some directional momentariness of movement.'

Also, in plasma physics, the word-sense **helicity** is applied to quantify various aspects of a magnetic field structure or the extent to which a magnetic field wraps around itself as a representing unit.

→But, from the standpoint of energy, exceptions are that the energy elements are **intense** or **extremely** 'versatile and **indeterminable**' on the point of **spin and direction** in any assuming discrete unit-value primary mass-energy of 'IEfu' since on the aspects of energy elements and many other terms on 'energy physics' including the generable **energy-Times**, comparably in the baby stages of the Universe is much more flourished than the holdings /assets by the matured 2nd-energy-belt.

*(Here, '**energy-Times**' derive following the elementary levels functionalities of energy elements.)*

→Besides, on the aspects of **dilution and diversities**, energies in the matured 2nd-energy-belt are much more progressed and dynamic than the containing energies in the fecund-Universe since the matured 2nd-energy-belt occupied residual form of energies, so the holding energy elements or working essences proprietorially feeble due to, either of componential (that is, *individual status as energy-element)* total losses or weakening of performing effectiveness of energy elements whenever justify an amount of mass-energy as unit/s in the matured 2nd-energy-belt.

*(So, from the viewpoint of energy, specifically to clarify the standard observable sense that applies to '**dilution and diversity,**' it is reasonable to say that the words 'dilution*

and diversity' are some state of parametrically discernible strategy within the vast /infinite like constant areas of energies in the matured 2nd-energy-belt.)

Above and beyond, elementary levels generating times, that is, the elemental times or overall energy-times of any unit-value mass-energy in the matured 2nd-energy-belt are much faster since time-governing components like motion elements are comparably **free and speedy** in action than in the IEfu.

*[{("**Some analytical clarification.**"*
*Here, **'free and speedy'** in action is justified based on 'motion' since **'dilutions and diversities'** gradually enhanced following the advancement and choosing the elemental diversities of energy elements.*

*Also, enhancement of dilutions and diversities are due to the formations of most micro-level various states (*that is, from low as thin to highly dense*) of objects that are reasonable to say some levels of zero-status-stable mass-energy frames, admixed of different forms of energies and countering functionalities from others.*

*Or, to simplify the above and ease of understanding, it is to say that immediately upon forming the volumetrically dignifiable mass-energy frames or objects (irrespective of their loose or compactness, based on the levels of energies), consequentially acted on the property as senses like 'dilutions and diversities' from the earlier status in a span of energies, so the impact of overall fluctuation to the levels of energies in the **universally*** (or true invariable) ***constant** boundary of the matured 2nd-energy-belt.}*

*{(Additionally, it may hint following the energy-level technical aspects that the True harmless quality /nature's 'motion elements' in the Pure-energies are somewhat like the **active** features that begin to characterize immediately upon infection acted to any most-micro point on an*

assuming fine-fettle unit-value Pure-energy in the prospective IEsu.

(Here, the word ***active*** *is applied based on the standard observable sense that may be unalike to understand the energy level's function.)*

Thereon, with the growing process of the diversity of energy elements or working essences, which were once only one and uniform in all aspects of properties, the motion element and many other infected ***energy elements*** *(these* ***energy elements*** *mean 'constituent or deep-down the working essences' of any assuming unit-value's pure-status of Pure-energy) functionally begin to act in a disruptive manner from the earlier positioning features (Absoluteness-status) of Pure-energies.*

(Here, it may requote that the motion element has only a solitary feature than the other energy elements since its exclusiveness of consistent 'motion' regardless of metaphysical conception granted by the word Pure /firstborn, impure /adulterated, and Purest or finest status.)

But, eventually, the motion elements at their extremely refined or Purest state or ***Purest-energies*** *somehow could organize and manage to lock due to the roles of 'universal physics' in the 3rd-energy-belt of the matured Universe.*

(Philosophically, the ***'Purest- energy'*** *in the 3rd-energy-belt represents the godly features and heavenly faith like the Soul of the actual Universe since technically, the 3rd-energy-belt performs as the honest and solo stabilizer that performs to hold the matured Universe at a fixed and unchanging position providing the dignity as eternity in all aspects, in the womb of mother Supreme - Universe.)}]*

→Also, following the status of **'dilution and diversities,'** the earlier mentioned point gives hints concerning easy determinable **'spin and direction'** of objects, objectual particles, or microparticles in some conditionally suited 'levels like grades, types, classes, and sub-classes of energies' through the naturally evolved computation systems of intelligent living species, and in addition, in reliance of many non-natural creations as measurement tools that plausible directly or indirectly *(that means, following some ways of analyzation)* to an individual or their group assessments while existing in or travel through the suitable atmosphere.

*{(Now, somewhat may be exemplified by assuming easily discernible **spin and direction** of an observable item due to its rotations following some standard or conventional magnitude and directions in some low resistive surroundings or circumstances whether regarded as objects or energies.*

*(Indeed, here, the observer or their group must be intellectually aware of the technicality of the words **'spin and direction'** and whether the in-built measuring system as a living being, Artificial intelligence, or alien is directly able to accept based on pre-set values or indirectly fed for acceptance after analyzing through different means.)*
So, for subatomic particles, the senses of the word 'spin and direction' are somewhat unconventional than the observable standard conception of intelligent human species.

*But energy levels of 'energy-elements' that sense **'spin and directions'** are even far and will seemingly be secluded and illusive when judgment is performed based on the prototype classical and the sub-atomic levels accepting reasonable techniques.)}*

→ [{("Some analogous elucidations of several objectual words that are employed for convenience to grasp energy level operations through the standardness of intellectual conception."

*Some earlier mentioned word like **indeterminable** is applied based on the terms of some, that is, delimitation of dimensional discernment or measurement.*

*Also, the words **intensity** or **extremity** on the term of versatile roles and determinability, which delimits to some extent the advanced stages of mass-energy, mass-energy frames, and objects since the sense 'logic' based on the degree of order and disorder or entropy, intellectually confined by some measurable reasoning, assess based on some accessible tools or systems of locale in a native nature of the 2nd-energy-belt.*

→Dilution and diversity

*Further, the standard objectual sense '**dilution**' means decreasing the **concentration** (*predominantly devoted to solution*), and '**diversity**' is demonstrated as multiformity, multiplicity, heterogeneity, and likewise.*

*Also, on the aspects of word-sense '**concentration**,' which determines the condition or level of '**dilution**,' predominantly applies to describe various states of matter like solid, liquid, gaseous, and plasma, since proprietorially, each of the four states of matter and their sub-states has multiformity on the point or term of distinctiveness.*

{But it may well be agreeable to everyone (regardless of theist or atheist) *that all those multiform state-of-matter follow the 'law of energy conversion' that somehow assures their (that is,* reasonable types of energies*) reappearance epoch after epoch, which allows one of the most fundamental levels of comfortness for existence as living species for generations like human beings, besides making comfortable strategy for survival in a place of habitation.}*

{(So, from the standpoint of energy, it is to point out that utilization of those objectual word senses, ***'dilution and diversity'*** *and so on, seemingly be remote to understand.*

Since the 'most-momentary variableness' saturated (that is, flooded) throughout the coverage of the matured 2nd-energy-belt, retaining better implication of words ***'dilution and diversity'*** *for objectual purposes, like the easy discernible techniques that developed a natural way to intelligent living systems of bio-species in this and many non-bio or alien habitable natures maintaining the dilemmatic sense 'nothing' or 'vacant' as hidden constant for measuring consolation.)*

(Further, the above clarification may be simplifying a little, following the standardness of objectual conception about the classical products, which encounter a 'computational space' like nothing or the vacuum to perform their actual most momentary performance for all sorts of outcomes related to the sense-like ***'dilution and diversity,'*** *maintaining some parametric cognizance as acceptance, regardless of the energy levels consistent diversifying accountable micro momentary roles.*

Also, for the point of microparticles, modern conception has extended to wave-particle duality functioning by comparing the classically recognizing momentariness activities.)

(Whatever the trendy concept of intelligent beings finds to operate and analyze the attainable objects or particles following some 'mystery' or 'unknown' like constant in the matured 2nd-energy-belt.

But factually, energies are everywhere and factually fastest in the aspects of momentariness, and most multifarious that regarded as extreme diversities, whether it segregates by using the sense like types, grades, sub-grades, classes, sub-classes, and so on, which all distinguishing categories *can utilize for convenient, by employing an ordinary word like '**dilution**' based on some reasonable parametric values.)}*

→ *(Also, based on the dimensionally discernible intelligent beings in this biological nature, many of the employed objectually befitted word senses are likely to be remote, or at some points, it will appear reasonable for realization but seemingly inauthentic to apply for the Pure-energies.)*

→ {(***Preparedness of three energy belts.***

Now, for the infectious energies in the 'IEsu,' and subsequently the follow-on ***evaluation*** *as diluted or weakened formed of energies, gradually disseminated as a* ***preparedness*** *of three energy belts throughout the Universe.)*

(In general, or based on objectual sense, ***preparednes****s entails a continuous cycle of planning, organizing, training, equipping, exercising, evaluating, and taking corrective action for ready-made items.*

*But for the Universe and energies, the sense of '**preparedness**' implied facilitating comprehension of the consistently growing physics in the affected 'state-of-Absoluteness,' which* (physics) *in a complex multi-directional manner could organize everything within the limits of the womb of the mother Supreme - Universe for the avoidance of any factor of the 'consequence of physics' that may lead for exaggeration and sense like impact to the periphery between the baby-Universe and the Supreme - Universe, and preclude the infectious roles of energies besides, it* (periphery) *impede of spreading more and more limitless or infinite on the aspects of spaces or areas.)*

(The foremost primary stage's 'periphery between the Supreme - Universe and the IEsu ultimately transpired into an eternal constant for everything in the Universe.)

(For the nicety of the above, a drawing SU16 depicted to apprehend a minimum, following the perfection of some dimensional discernment of intelligent species to indicate how multi-directional ways of energy-values or Physics-values factors perform to give a perfect structural buoyant position of the Universe in the womb of the mother of the Supreme – Universe.)

The prospective periphery covers all-around to restrict the embryonic activities in the IEfu between the progressing stage of the baby Universe and the Supreme - Universe.

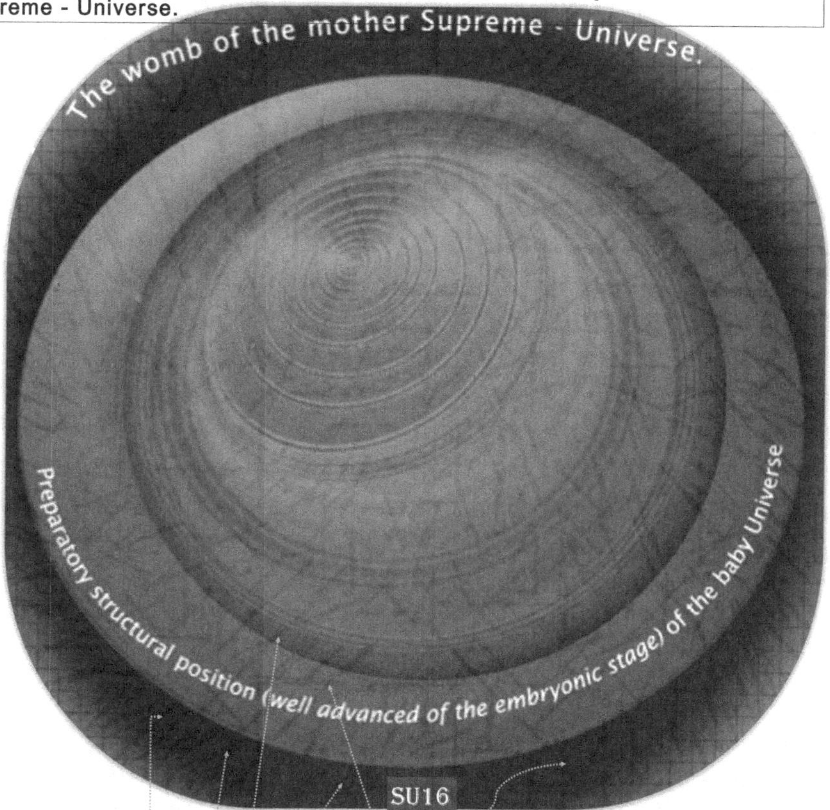

Multi-directional energy-values or Physics-values factors perform to give a perfect structural buoyant position of the Universe in the womb of the mother of the Supreme – Universe.

[{("Some analytical clarification."
'Evaluation and devaluation' determining the objectually applying sense 0.

*The word sense '**evaluation**' here may justify as 'disintegration and scatteredness' on the aspects of energy-*

elements from absolutely integrated Pure-status of Pure-energies, leaving no room like objectually applying sense as vacant, empty, or nothing, etc.)

*(A little more clarity about the sense of '**evaluation**' obeying the objectual insight like 'zero /0' is standardly accepted as a static value regardless of **own-self** mean respective status of a living being, and as recognizing any of individual state of matter, which distinguish follow-on their diversifying proprietorial features like solid, liquid, gaseous, and plasma, even though based on their specific terms, each of the states of matter has wide varieties like an iron ball, book, ice cube, etc., justified at their solid state.*

So, I find the symbol '0' is one of the better options to apply at energy levels constant for expressing the evolution and devaluation, observing the mathematical ways of computation and analyzation practices among the intelligent species in this habitable globe.)

*(One thing to quote on the aspects of the word '**own-self**' is that the individual position (*or the overall constitutional value of a physique*) of each unit-value living being tends to justify them as 0 like static-value or constant for convenience and satisfaction of assessable and measurable items.)}*

(Since '0' is a static /dormant state beyond the objectual sense and following the standard conception of living beings, I think, at this juncture, '0' and its manipulating strategy with decimal values, numerals, and some reasonable symbols will be the sound option to explain energy and energy level functionalities from the initiation of the Universe to became eternity with juvenile status.

For clarity regarding '0' and its multiway utilization, I refer to some drawings in this part and other parts of the book series where time-to-time tried to elaborate as appropriate.)}]

→ *(In fact, 'Pure-energies' of the Supreme - Universe Truly signifies the Singularity and 'abs0^∞' featuring, preserving all imaginable, probable, and any unimaginable working essences intact at a uniform state, and its status is far-far beyond to realize through the specific way evolved intelligent beings in this nature.)}]*

iii. Wreath around.

Considering any unit-value energy of the 2nd-energy-belt, the overall elemental outcome as 'elemental values' tends to **wreath around** to represent the role of '**helicity-energy**,' which somewhat performs as a boundary of the unit-value when the constituent energy elements integrally engaged in regulating the internal physics of the unit-value in the conduciveness of surrounding atmospheric-Physics.

→Besides, a **conducive atmosphere** assures appearance, changeover, or following the objectual sense may justify as desublimation of energy to make a discrete amount of the early phase's mass-energy, thereupon mass-energy frame and later stage's objects, and so on recurrently appearing energy products following the relentless energy levels physics.

Also, it (*a **conducive atmosphere***) gives some levels of security to all energy products, reserving the inter-tangling relations with the 'natural physics' of locale and the 'universal-Physics,' which all together in a coordinated manner rules throughout the matured 2nd-energy-belt.

→**Some talk about natural laws and universal laws.**

Therefore, in a remote sense on energy point, it is to speak that violation of parametric limitation of the **natural laws** that regulate all-natural way habitating objects in concurrence with the natural -Physics led to early or first corrosion from a state of standardness since on the multidimensional standpoint, universal-Physics is inter-tangled throughout the 2nd-energy-belt.

{(Technically, following the perspective of energy, the conception and application of ***'natural laws'*** *do not mean that it will restrict to some limitation of adaptability like this biologically evolved nature /Globe that supports only specific types of living species.*

But the formation, regulation, and time-to-time amendations of all ***'natural laws'*** *directly or indirectly intertangled with the universal laws, assuring the full-fledged appearances of all the assets of the matured 2nd-energy-belt forever, regardless of whether they (assets) applicable or not for versatile functioning in the reasonable* ***atmospheric physics****, that is, the physics which may arise due to the reasoning of* ***'Twig-atmosphere and energy-Atmosphere.****'*

(Besides, applying the philosophical sense, it is to quote that nothing could transgress the ***universal laws*** ('universal laws' derive and regulate all sorts of physics) *and interrupt the perpetual systematic works of the matured 2nd-energy-belt, which assures the eternal juvenility of the actual Universe within the mother Supreme - Universe.)}*

→{(*Here, technically, one thing to mention for conceptional clearance is that energies in the IEsu zone or the pre-embryonic areas of the Universe incrementally get diluted or weakened, or 'disintegrated and scattered' on the aspects of energy-elements from the earlier standing of absolutely integrated Pure status of Pure-energies within a fixed boundary or periphery that covers some areas or space in the womb of the Supreme - Universe, which entirely negates the objectually and observably demonstrated* ***elastic property*** *for energies.*)

(In common, the objectual ***elastic property*** *is something like 'expansion and contraction' in the reasonable items, or 'nothing' like mysterious constancies.)*

So, based on the three-dimensional notion of intelligent beings in this bio-Globe, the minimum way may try to convince, elemental displacements and elemental gatherings of energy-elements from the initiation to reaching the mature stage of the Universe, ensure the intactness of total space or area occupied by its (actual Universe) *birth in the Supreme - Universe.)}*

→Objectually, the sense **'wreath around'** assesses and recognizes following some imagining or discernible areas or items.

But, from the standpoint of energy, technically, the sense **wreath around** is quite unlike the observably recognizing sense, and at the energy level, wreathing performs multi-directionally whether referable or not-referable as a unit-value mass-energy frame or object and where there are the appropriate (*that is, conditional*) circumstances following some degree of lower or upper phases energies all around the conditional limits.

Further, for ease of comprehension, the drawing SU15 is a dimensionally restricted aerial view presented with four individual sets indicating the overall elemental outcome as 'elemental values or physics values creations that tend to **wreath around** as an elegance of **'helicity-energy**,' which somewhat performs as a boundary of the unit value when the constituent energy elements engrossingly regulate the internal physics of the unit-value mass-energy instantaneous of reasonability of surrounding supports.

iv. [{("*Some analytical clarification.*"

Here, one thing philosophically intends to quote that on the point of functionality, and based on the assessable parametric ranges from a possible lowest to the highest degree on the perspective of the human species in this biological Globe.

Analogously, energy levels work in the baby stages of the Universe are ***incredibly prompt and assertive*** *than the standard senses apply to recognize objects, objectual products, and diversified classes of energies on comparable strategies following the standardness of natural ways evolved intelligent living system and their non-natural creations for assessment of accessible items in the matured 2nd-energy-belt.)*

*{(By the way, I think the sense '****incredibly prompt and assertive****' and many likewise objectually recognizing pre-set words is reasonable to apply for the energies in the baby and mature stages of the Universe.*

*However, in this day and age, based on the modern intelligent species in this bio-Globe, the same ('***incredibly prompt and assertive***') may be denoted as* ***'heavenly spooky actions'*** *since intellectual species, constitutionally* (that is, on volumetric aspects or structural value of any physique) *restrict by some degree of directions for general analyzation that developed following the natural laws of habitability.)}*

→ *{(Besides, following the diversity of energy products, sub-products, and energies, it is to quote that functional role (functional role at this stage may mean causal functions) due to the* ***momentariness*** *of motion elements are incredibly faster in the 2nd-energy-belt than other energy belts and the baby stage of the Universe, except the extremity on the point of 'kineticism and neutrality' of the Purest-energies in the 3rd-energy-belt.*

This incredibly fast momentariness of motion elements in the matured 2nd-energy-belt is due to the insufficient or diluted quality of energy on the aspects of originalness and numerousness or multitudinousness of energy elements.)}

*(One thing to quote is that the **momentariness** of motion elements in the 'energy -Ocean' of the 2nd-energy-belt is not the motion elements' self-bearing property. Instead, **momentariness** is a causal effect due to the diversifying roles of other energy elements in any assuming mass-energy, mass-energy frame, and object, including their constituent motion element. However, in principle, the pure or some state of the finest forms of motion elements tend to flow uninterruptedly.*

*So, applying the observable sense, this diluted quality of energies may mean something like a **free space** that works as one of the comparable factors for faster roles of motion elements and other energy elements.*

*(Here to quote, on the aspect of energy, no word sense can apply as a '**free**' concerning in any space or vacuum-like idea.)*

*(By the way may speak, following the observable sense, the word '**momentariness**' is logically introduced for the application of energy levels functionalities as a 'tinniest' the most fleeting /momentary value, whether acceptable through the standardness of conceptional limits that evolved as pre-set value to intelligent living species in the course of natural practices, or beyond of their imagination.)}]*

v. In drawing SU15, it assumed some units of 'mass-energy' stepwise occupy one, two, three, and possibly more elemental positions following the most fundamental or common one, the **motion element**.

This **motion element** represented as base depicted as 'c, 0 or somewhat like the sense as constant,' and on a functional perspective, it (*motion element*) **somehow** (*at* *this stage, the word sense* ***'somehow'*** *justified in 'multi-dimensional perspective'*) covers **almost** throughout, in any form of imaginable or beyond most-micro to massive discrete unit-value mass-energy that determines no place of vacuum or nothing like ideas.

(Here, the word ***'almost'*** *is implied based on some dimensional logicality that, in general, is accepted by the intelligent species in this habitable globe. Simple, like an observable stationary or non-moving term, aggregately granted as no-motion or the same at energy levels, it is considered 'no roles of constituent motion elements in the item.'*

Otherwise, the actuality of the motion element's 'neutral-motional' propriety is intrinsical, and its existence is everywhere, which is exclusive from the other innumerable types and features of energy elements.)

→So, from the perspective of the sense larger like 'infinite,' for the ostensive 'Purest-energy,' the most common sense as 0-symbol or formulaic purposes used alphabet 'c' (*the peakest speed*) at this stage, fundamentally may justify for application as the foundational role and everlasting constant of the 2nd-energy-belt to ease of clarification and understanding that may match perfectly for the intelligent living systems in this biologically evolved nature /Globe and possibly some other alien intelligent species.

→By the way, briefly, the energy elements or energy levels' working components are possible to discern following their proprietorial roles, that is, based on the aspect of distinguishing functionalities.

Or conveniently, the same that is, 'energy elements' may distinguish degree-wise following their distinctive attributions, which (*that is, degree-wise characterizing)* is somewhat a standard conception insinuated by the intelligent living species for recognition of energy products distinctively.

→Some hints about the idea of 'influencing-energy-element (Iee).'
Further, at this stage, assuming the motion element proprietorially acts as a base like '0 or the alphabet 'c.'

So, based on the standard numerical essence or value of '0,' scale-wise figure '1' will be the closest to 0.

{(*Here, '0' traditionally signifies a static value (*static value is not truly 'nil or nothing'*) /status or a constant/unchanging factor like* '**c**.' *But from the perspective of energy, '0' can again be 'evaluated or devaluated' by utilizing the directionally determining some levels of positive and negative of the decimal values notion, besides by employing some appropriate numerical values and symbols, reasonably indicated in the drawing SU17.)*}

Or, the same may be defined assuming ‘0’ represents the place (c), or say, motion deriving essence (c) or 0-value symbolizes a constant or some state of the unchanging factor for deriving speed bearing dynamical property.

Therefore, based on the innumerable status of energy elements, figure '1' (*that is, a mathematical representation*) can be denoted as the **'influencing-energy-element (Iee)'** since 'degree-wise' and 'characteristically' it (**Iee**) primarily function as swaying roles in unit-value energy, mass-energy, mass-energy frame or object.

→ *[{(**"Some analytical clarification."***
*By the way, for ease and conceptional clearance, one thing may quote following the perspective of a variousness of energies on the aspects of quality, the proprietorial essence of **influencing-energy-element (Iee)** may again be sectioned and sub-sectioned by using methodically*

*applying decimal values to some limits of standardness or logic limits of intelligent species (*depicted in drawing SU17)}.

→ *{(****Utilization of naturally accepted word sense 'free-flowing.****'*
Further, in the 2nd-energy-belt, it may quote for conceptional clearance concerning proprietorial aspects of the ***free-flowing*** *energies, which are entirely unlike the mass-energy, mass-energy frame, and then the advanced stages energy product-like object/s since, whenever some amount of* ***free-flowing*** *energies encase or congeal by some areas (*that is, parametric stance), *their constitutional physics immediately get restrained* ***to preserve the integrity*** *of the internal assets or working essence or energy elements following the roles of surrounding atmospheric-Physics.*

*By the way, it is to quote that a trend like '****preserving the integrity****' of any unit value mass-energy, mass-energy frame, and object is not a mysterious practice. Instead, this practice is the* ***gift*** *from the mother Supreme - Universe endured by the birth of the Universe.*

*{(Here, I think the word sense '****gift****' is befitting to apply as 'some form of legacy /heritage' since in the mother Supreme - Universe, all the imaginable and unimaginable componential components are at one and only one uniform state, that is, state-of-Absoluteness.'*

*So, upon the birth of the Universe, any of its componentially bearing propriety and related factors are the gift, which was once only a uniform or one-state, which can symbolize as 'abs0^∞' (*this 'abs0^∞ was clarified earlier also as appropriate, in the book series*).}*

(Also, a philosophical viewpoint may speak that immediately upon getting the individuality as a unit-value mass-energy, mass-energy frame, or object regardless of

relative status from most-micro to massive structural value (that is, area coverage*), it loses the* **free-flowing** *roles and has to tackle /confront a wide-ranging external Physics in the energy -Ocean of the 2nd-energy-belt.*

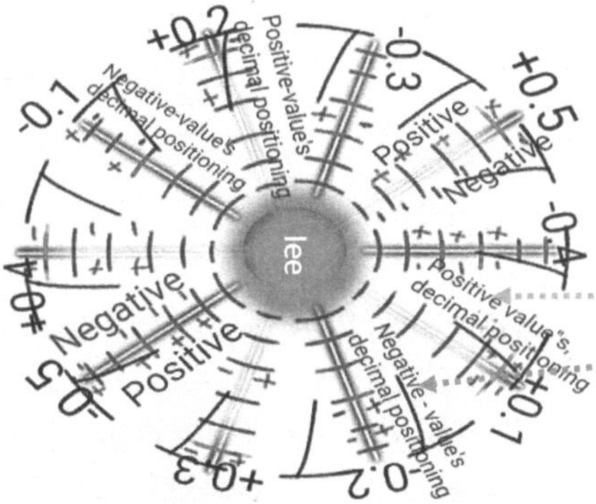

1- Simple conception that standardly follows two directions, maintaining an indirect neutral value or line as acting constant for positive and negative ways of directional measurements.

SU17

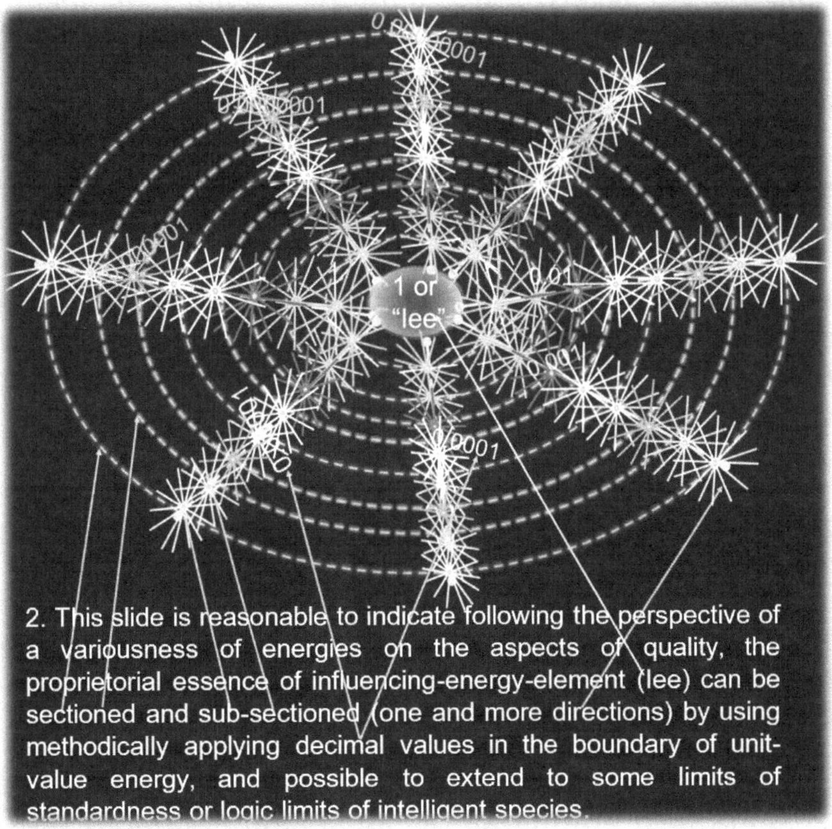

Here, a little more may clarify the observably applied word sense ***free-flow****, which is quite different and seemingly unrealistic when attempting to understand the energy levels terms as* ***free-flow*** *since the functional part of the sense 'momentariness' is extreme and beyond any comparable mean that is generally appropriate to grasp /understand by the standardness of living beings' conceptions.*

Standardly, living beings' conception accepts the sense ***'free-flow'*** *as the movement of anything without stopping or continuous in a natural way, which is determined through the built-in measurement system restricting some realizable constancies, whereas hiding or averaging out the reality of affectable many deep-down (means elementary levels) of energy levels momentary functionalities that are also accountable in terms of actual conditionalities of any accessible and assessable items.*

From the above, it may be convincible to intelligent beings in this biological nature /Globe that the objectually presented words 'free' and terms like ***free-flow****, free will, and so on senses are restricted or restrained by some reasonable conditionalities. However, at energy levels, momentary roles do not allow anything to be accomplished freely, following the interlinking relationship and the universal laws of the 2nd-energy-belt.)}]*

3. The objectual word sense 'string' and its energy levels utilization may fulfill the term' as energy-string.
In drawing SU15, the word 'string' is generally employed to match a standard observable perception of living beings that satisfies the purposes of tying by using fiber /cord /line and likewise which fabricates from some forms of sensible matter, material, etc.

Even the same may link with the remotely formed **psychological simulates** (*or emotional creations*) for tie-up connections or possibilities of maintaining interlinking relations to nearer and unknown distances in terms of functionalities.

{*(Here, out of many examples, which is possible to demonstrate in different ways, however here is the one something like the word sense '**blood relation**' is a topic of logical family connection nearer or a far distance from the parent. Indeed, this logical value's **blood connection** can extend directly or indirectly when* **psychologically simulates** *exercise beyond the limits of **biological**ly evolved human beings (*here, **biological** means one specific class of life-orienting energy flourished in this bio-natural earth.*)*}

→Although the above objectual senses related to the word '**string**' are remote to associate with energy.

However, somehow analogously, the same word '**string**' attempts to equate with energy levels characterization denoting as the word **'energy-String**,' **energy lines /cords,** or more deep-down at elementary levels as elemental values' thin rapidly pliable threadlike **elemental-lines** *(indicated in the drawing SU19)* that functionally perform as ties in a relationship with the most quickly active energy elements in unit-value energy, mass-energy, and mass-energy frame or object, and even when they (*that is mass-energy frames or objects*) are independent, someways reserved interlinking relation directly or indirectly to near and possibly a far to observably isolated distant.

(F*ormation of these elemental lines, **energy lines, overall may justify as 'energy -Strings'** fabricate from some form of **'energy signals'** when there is diligence of some elementary levels or energy levels activities, and its significant utilization remains intact in the matured 2nd-energy-belt.*)

i. Technically, from the standpoint of energy and following universal physics, the **causal functions** determine the possible direct or indirect interlinking relations between the units of mass-energy, mass-energy frames, and object near and far distance using some form of **energy signals** or **strings** to maintain near or distance ties.

*(Here, from the viewpoint of energy, **'causal function'** may mean the interlinking input-output relationship regardless of objectually distinctive characterization that follows mainly the past and present but may be known, unknown, or some possible future sequences.)*

→The sense of **'causal function'** may somewhat exemplify following an observable sense about some balls floating in a calm pool of water. Under this calm circumstance, the surrounding atmosphere must be **parametrically reasonable** to maintain the tranquility of the water of the pool and the balls.

Or, implying some energy level means, it may speak that the **parametrically reasonable** tranquil status overall represents a computational zero to an observer-like intelligent being regardless of how accountable the energy level consistent causalities. *(Or the same may simplify and say that an observer or intelligent being own-self characteristically acts as '0,' which is a satisfactory value for the comfortness of assessable measurements.)*

Whatsoever the observable sense discerns, from the standpoint of energy factually means, in some specific areas, all the containing mass energies, regardless of transiting or local born representing mass-energy frames, objects like air molecules, balls, the pool, and waters, including their consistently generating physics are interconnected to maintain balance co-ordinately by mean of producing some sorts of remote energy signals (*energy-strings*) or objectively recognizing cords /strings for connecting as interlinking relations of that specific area for maintenance of tranquility in the **parametrically ascertainable area**/s.

{(Based on the objectual sense, the earlier mentioned point as a ***'parametrically ascertainable area'*** *is generally determined, following some limits of averaging* ***physics*** *values submission like temperature, pressure, resistance, force, etc.*

This ***'physics'*** *derives due to the constituent qualitative energy-elements or working essences' incessant momentary activities in any objectually consolidated position since energy levels' momentarily raising many energy levels physics or energy Physics remain indetectable but reasonable for that* ***'parametrically ascertainable area'*** *to maintain tranquility status of the containing.)*

(By the by, from the standpoint of energy, one thing to recite for convenience of concept, the word sense moment is restricted by time, energy -Times, or elemental -Times.

So, the perceptibly standard term 'times' factually engenders from the deep-down to the elementary levels of an amount of mass-energy unit, and the elementary levels momentariness led to some form of physics values derivement regardless of discernible or not, following the standardness of measurable techniques of biologically evolved intelligent beings, and the discovering aiding non-natural mechanism for discernments or measurements.)}

Now, suppose the water of the pool is disturbed by some means or dropping an extra 'ball,' the ball's spawned physics (*that is, the external physics implied by using an extra ball*) will impact multi-directionally from water, atmosphere, and then causally will **tend** to relay through the faster-forging **energy-strings**, all-around the co-ordinately maintaining tranquil areas *(attempted to depict comprehensibly with the drawing SU18*).

(H*ere, the use of the word sense '****tend****s' to indicate uncertainty about the external physics values implications.*)

ii. Also, note that any form of mass-energy and mass-energy frames or objects consistently release some form of energy regardless of proprietorial characteristics such as physics, energy signals, and so on due to the deep-down at elementary levels non-stop activities.

And, as a unit-value energy, mass-energy, or mass-energy frame, all internally emanating releases (*as energy forms*) at the periphery carry lots of elementary levels of in-depth functional entities, which are most important for the existence of the unit/s itself, besides spawning **information** when there is the reasonable or suitableness of atmosphere that allows maintaining the self-ness as a state-of-stability.

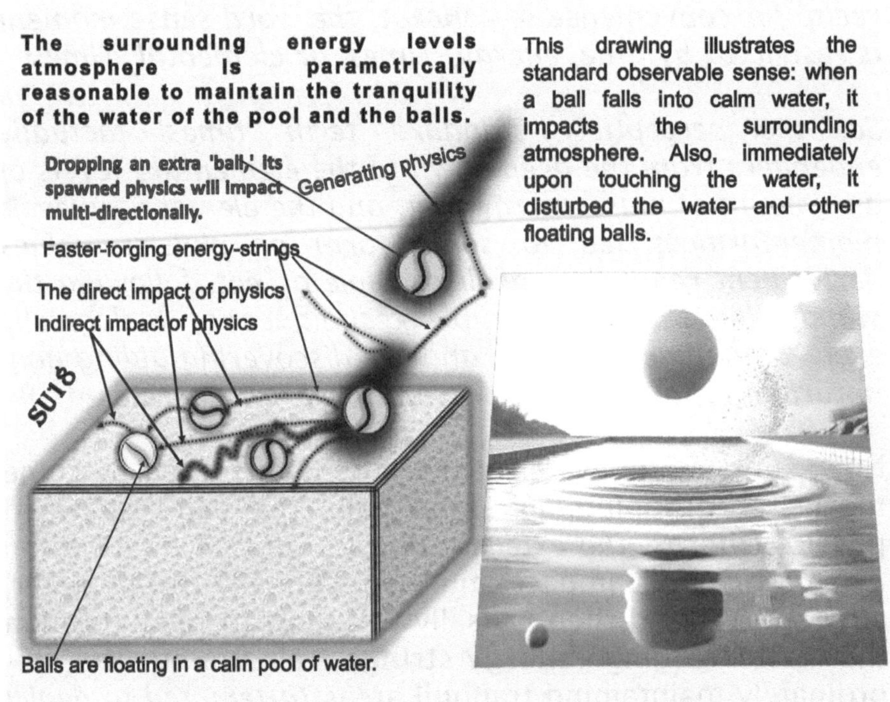

*(The word sense '**information**' is employed to indicate some format /spectrum of energy signals that expose consistently through the periphery of mass-energy, mass-*

energy frame, or object, about the internal momentarily roles of elemental physics of constituent components in a surrounding circumstance.)

→The in-depth elementary levels 'energy lines' or 'energy signals' maintain or hold isolated /personal relations of constituent energy elements in any unit-value mass-energy, mass-energy frame, object, and objectual products under **conduciveness** of surroundings near and unimaginable **far-distant** on the perspective of **multidimensional functionalities**, which is entirely unlike the observable delimitations that restrict intelligent species in many aspects to understand the factuality of energy levels works. *(Also, the sense of **far-distant** can apply to small and big units and the interior of a unit-value huge mass and complex arrangements.)*

{(*Here, at energy levels, I think the word '**conduciveness**' is appropriate to explain the planning and organizing an amount of energy in the infinite areas of 'energy-Ocean' to make a mass-energy, mass-energy frame, and successively the object following the various levels of physics or energy-Physics that are consistent with some form of reasonable or outlandish consequences.*)

*(By the way, for clarification of '**multidimensional functionalities**' on technical aspects, one thing may cite that the 'ostensive Purest-energy (or soul)' acts as an unconditional 'universal-Constant' that covers the nook and cranny of the 2nd-energy-belt, which functionally represents the fastest and neutral media for all observable or overt and non-observable or covert activities going on forever throughout the 2nd-energy-belt.*

For example, on the above point, it easily can be indicated towards the cobwebs, a single thread spun, which acts as a constant and media to inform activities. So, if there are any reasonable covert and overt activities at any point in the thread, it could span throughout the coverage of the cobwebs.

Indeed, the standard cobweb works on 3D (three-dimensional) *tracks and is capable of handling /regulating some goals compared to something having more dimensions, but entirely unlike the multi-dimensional functioning abilities of the 2nd-energy-belt.)}*

→Although the objectually presented word sense '**string**' will seemingly be inappropriate to apply to multi-directionally performing the energy levels' most brisk, delicate, and sensitive works. However, its (*string*) proprietorial attributions are reasonable to employ to convince in-depth of energy levels tie-up relations.

Also, it seems works of string are reasonable to express on three-dimensional pursuits, which in some ways can be utilized in many elementary levels' purposes, besides foundational networking arrangement of very fast performing connectivity of energy elements or working components like 'Motional-string for motion element,' 'temperature-string for temperature element,' pressure-string for pressure element, force-string for force element, and so on many momentarily altering components, which directly or indirectly act as interlinking **agent**s maintained or hold energy levels all forms of divergence relationship throughout the matured 2nd-energy-belt. *(Here, the word '**agent**' technically may mean some levels of the spectrum of energy signals.)*

Whatsoever, I think the sense of '**string**' is appropriate to employ from the progesterone stages of the fecund-Universe (*IEfu*) immediately upon the formation of unique or mysterious modules that release and deep-laid or lose track of during unusual symmetric and asymmetric activity among the break-off units of infected or affected

Pure-energies in the pre-embryonic areas for the progression of progesterone in the Supreme - Universe.

However, the sense 'string' is inappropriate for the 3rd-energy-belt since it occupied only 'Purest-energy' and is a True single or mono-state property. Also, there's no possibility of raising questions about any form of diversity besides it (*Purest-energy*) does not require any benefit from interlinking agents like string.

(By the way, one thing is to remember for convenience the standardness of intelligent conception that, at energy levels, there are no such observable words appropriate to apply as vacuum, nothing, or empty since some forms of energy are reasonably efficacious, entirely non-functional or neutral on the comparable strategies with assessable surroundings.)

iii. {("*Some analytical clarification.*"

The drawing (SU19) is a depiction to hint at some of the earlier-mentioned words and their utilization for the functioning of the deep-down of unit-value energy.

(Firstly, one thing to requote for the convenience of conception is that the total amount or all the energies in the vast 'energy -Ocean' of the matured 2nd-energy-belt cannot sublimate into mass-energy, mass-energy frames or objects since, after the post-stage of development, 'ostensive Purest-energy' gets elevated more and more due to consistent congelation process among the types, grades, classes, sub-classes, and so on of ***residuum*** *form of energies than the pre-mature state of the most tumultuous 2nd-energy-belt, which undermine the availability of the appropriate form of energy components or elements for forming of any fresh form mass-energy but the only option of recycling ways of appearance to become mass-energy frames and objects.)*

(Since 'energy elements' are engaged consistently in maintaining individuality as status, some form elementary phases 'signals of energy elements,' which are the most fundamental levels of initiating 'energy string,' fleetingly create and perform the fastest and most versatile functions within a mass-energy or mass-energy frame.)

*(Also, briefly may touch on the sense of '**residuum**' to indicate that from the time upon the birth of the Universe and gradually towards progression with distinctive energy belts and their different peripheries, sparing with lots of energy components or energy elements, so there is no more originality of Pure status of Pure-energies and in successions, the 2nd-energy-belt behold largely, the feeble or low potency discharged from 1st-energy-belt and the 3rd-energy-belt.)*

→In drawing SU19, mainly the two points are marked with 'a' and 'b' only to indicate some of the componential standings: -

*a. '0 or c' represents a base or center position for 'motion-element,' and then the possible extension to some areas following **decimal** numbers for placements of fleetingly changing types of other energy elements.*

(Here, **decimal** quantitative is utilized for ease of understanding since it became a mindset derivative to intelligent beings in this biological Globe.*)*

b. Internally confined networking arrangement of energies in unit-value mass or a mass-energy frame, this energy networking facilitates regulation of all the constitutionally set energy elements, straightly or subordinately to call for any change or to be at the standard of the unit-value energy under the conditional surroundings.)}

a. '0 or c' represents a base or center position for 'motion-element,' and then the possible extension to some areas following decimal numbers for placements of fleetingly changing types of other energy elements.

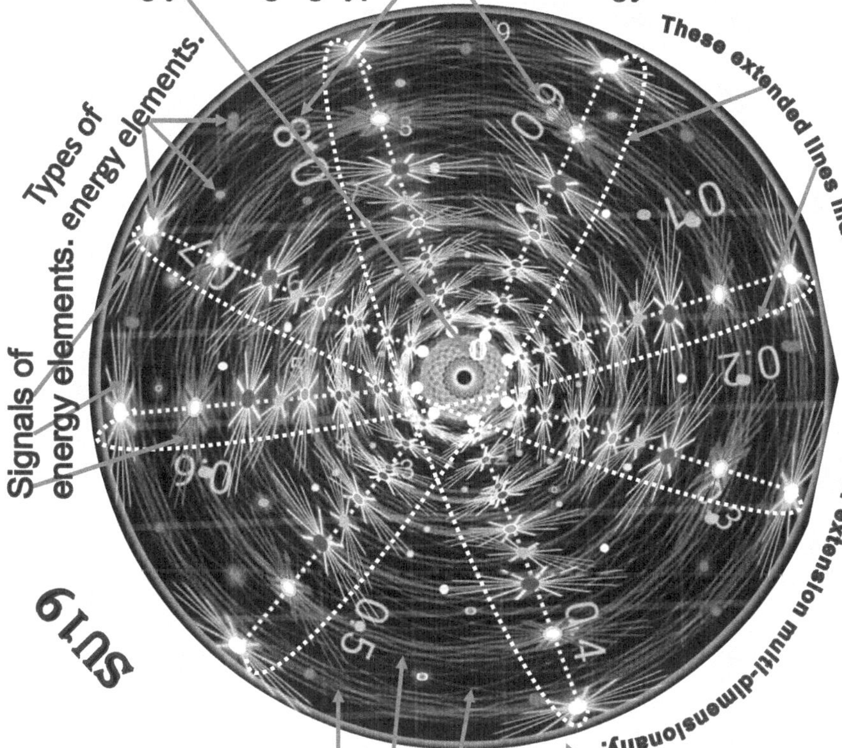

b. Internally confined networking arrangement of energies in unit-value mass or a mass-energy frame, this energy networking facilitates regulation of all the constitutionally set energy elements, straightly or subordinately to call for any change or to be at the standard of the unit-value energy under the conditional surroundings.

4. 'Some notes concerning the features of Generic-energy and Genomic-object or Genomic-matter etc.'

It may be known to all that the word senses generic (*means relating to, or belonging to a genus)* and genomic *(means the study of all of an organism's genes, interactions among genes, and the roles of the environments in affecting them)*, which standardly apply on physiological terms for livings in this biological globe.

Indeed, it may be opinionated to comprehend with ease when applying those physiological word terms (*genomic, generic, and likewise*) and directly related to energy level functionalities. Anyway, at this moment, I believe using these familiar terms is more effective in persuading modern intelligent individuals rather than introducing new words that could make understanding more difficult.

i. From the perspective of unit-value energy, pellucidly may be defined as the initiation of differential values that lead or incline for 'equilibriums' due to the inherent or genetic discipline /domain for the absolute blueprint or prototype stance from the Pure-energies of the Supreme Universe.

For the Universe, seceded units of energies from the pure status of Pure-energies means losing and gaining from a state of completeness (*that is, Truly Pure status*), which is the cause for the derivement of differences that lead to some lowest level of commotions and consequential outcomes.

And thereon tend for equilibrium or attempt to bring into equipoise, derive by some sorts of the novelty features of Generic-energy, the likeness of pure /vestal state of originalness inlaid as a blueprint to all offshoot energies.

→Further, any state of unit-value energy in the matured 2nd-energy-belt and its belongings like differential '**2b-energies** spatial /spaces' and there-from advanced stages various forms of recognizing matter-energies, matter-elements, matter-products, product-elements, product-matters, product-energies, the Cosmical-Universes, galaxies, planetary systems, then diversified living and non-living natures, and so on many versatile subsistent metamorphosed products.

All of those natively wangled or acquired belonging *(here, belonging means small or big units of 2b-energies)* share **'Generic-energy-features'** blessed from the Pure-state of Pure-energies regardless of massive and quantum or most-micro values identities.

→But, based on the objectual sense (*units of* ***2b-energies***), genomic exercises and demeanors comparably will differ from weakest to most decisive, or most increased leverage than the unit-value energy of embryonic stages of the Universe since percentage-wise entire amount of energies in the 'IEfu' shared by the three energy-belts without disturbing the **blueprint** of **Generic-energy-features,** that is like the genetic /ingrained trend to stay equipoise /symmetry somewhat replicating feature of Absoluteness-status of any imaginable unit-value Pure-energy.

(*The above quoted,* ***2b-energies*** *means the entire possessing energies in the 2nd-energy-belt.*)

ii. Also, following the objectual Senese, In the matured 2nd-energy-belt, the advanced stages of functionalities of Generic energies vary from tangential to egresses, or the same may be simplified by applying standard sense like 'realizable through physiological experiment' or 'unreachable due to egressed following massive and unanalyzable mutational changes.'

These Generic-energies features may be able to enumerate pursuing the quality and quantity of contented energy elements or working components in unit value objectual products.

→Hence, many naturally borne novel **Generic - hallmarks** that standardly characterize depending on genomic exercises of the feature/s of 'Genomic-matter' mean, an indirect link to Generic-energy-features.

Whatever, for ease of comprehension about the **generic hallmarks** and **genomic exercises** of the features of the **genomic - matter**. Abstractly, it may lay an example following the comparable senses among the units of individual bio-super-beings in this bio-nature and say that, as the ancestral features of a genus /species, all inherently apportion in a genomic manner to their descendants like modern species in this biologically habitable nature.

Due to the genomic exercises, each **bio-genus /species** contemplated with a **generic hallmark** and the natively or natural way on a time-scale allotted or obtained individual features that may justify as **genomic-matter-feature**.

iii. It is to quote that any unit of biological living species is reasonably justified as built by a simple or combination of multiple molecules from their tiny to reasonable in the shape of specific formats as a mono-state molecule *(like Amoeba and some others),* or their union as a group form organ with **comprehensive values** living form of energies.

The **comprehensive values'** of 'quality and quantity' of living forms of energies (*somewhat like the life-orienting biological energies for biological species*) may mean the limits of an outlining parameter.

So, parametrically limiting bio-energy, which naturally engaged for designing the bio-super-being species and others in this biological habitable Globe, cannot be considered perfectly equal to the super-livings inhabiting distant biological or non-biological nature/s of liveable globes in different cosmical - Universes.

However, they must be relevant to characterize as objectually indicative sense 'equitable,' apportioning with the generic energy features as living functionalities.

→Besides, it is reasonable to say that the bio-living species may formulate from mono as simple to poly like a multiplex unit of molecular organ, founded on the inbuilt quality and quantity of bio-energy sources that are obtainable mainly in a life-regulating adaptable atmosphere like this biological globe.

Also, equating with the unit-value Pure-energy's absoluteness status, the disposition towards equilibrium of differences is because of variousness. Accordingly, the trend for balancing the imbalance of unit-value energies in the 2nd-energy-belt became a **genomic exercise** that bear immediately upon the commencement of the Universe.

→The enactment of balancement is restrained /controlled mainly by the energy -Times or in-depth elemental-Times of energy elements due to the enduring active role of motion-value of motion-element that constitute deeply inside any unit-value energy, mass-energy, mass-energy frame or object in the matured 2nd-energy-belt.

→Here, to quote that some form of energy value emanates due to the consistent derivement of the Motional value of motion-element, which has the invariant attitude for devaluating and evaluating from an earlier status of unit-value energies, regardless of the standard conception of human beings' recognition as living and non-living energy products.

So, the attitude toward devaluation and evaluation is consistent with any differentiable unit value irrespective of energies or objects, and the tendency for balancement derives from the 'genomic blueprint' that carries from far-off absoluteness-status of Pure-energies.

iv. Logically, the feature of 'genomic energy' and the tendency to make equilibrium of differential values may be realizable on some standardness or specific levels of ideological parameters derived in the intelligent beings' in-built analyzing systems, directly or indirectly through aiding tools.

→The former /earlier stage's 'generic energies' flow in diverse manners that gradually enhanced transmogrification to their advanced stage, and thus, many energy factors evolved as competent and reoriented into genomic - energy even though their mass-energy frame participate for recognizing as different states of matter.

Thus, based on the standardness of ideologies to the living beings in this bio-globe, original energy features from their Pure stage, the embryonic stage of the Universe, and the advanced stages of diversified forms of belongings in the 2nd-energy-belt will be seemingly isolated considering the performance and properties.

Also, advanced stages' generic - energies usually carry through genomic bearers relying on the diversifying activities of objectual-energies and objectual-products irrespective of bio-grade and non-bio-grade energy release.

v. However, the most common thing of those un-specifiable Generic energy features and specifiable features of 'Genomic objectual or Genomic matter' are time-bound reckonable based on certain logical stances of positive and negative inspiration /influence /practices to intelligent species of various natures.

→It also may mean that the matured 2nd-energy-belt, Genomically characterizations are seemingly equal for specific grades of energies in a particular energy - atmosphere.

But, following the objectual sense, for the distant cosmical-Universes, those specific Genomical characteristics will not be identical other than the replica or remote of recognition.

It is due to the consistent transformation role among the innately set energy elements in the stable unit-value-energy or their union as mass-values energy, object, or like a living being, represented as a unit-value bio-mass energy frame.

Accordingly, intellect systems evolved with particularity may find it uncomfortable /illogic to link directly concerning to equate genomic feature/s in all respects with different or distant grades of objects and objectual-products, even though they carry the same gene of Pure-energies of the Supreme - Universe.

5. Based on energy, some fundamental issues and technical clarity on the factors of structuring the Universe.

Coordinating act of **elemental-value of energy-elements** (***energy-elements*** *are the utmost deep-seated functioning essences of any form of unit of energy*) to the unstable Pure-energies of the embryonic stage or in 'IEfu' co-efficiently derive for the structure of the Universe.

(Concerning the technicality of deep-down energy components like ***'energy elements' and their working essences, 'energy values'*** *are described frequently as appropriate in different parts of the book series.)*

The effectivity of acting differential value and tendency towards equilibrium among the detracted unit-value-energies ensures the consistent engendering and processing of known and mysterious energy elements, besides elemental values or physics values *(here, 'physics values' applied based on objectual sense*) diverse functionalities throughout the developing Universe.

Among the numerous working essences, some of the most significant energy - elements and their operative elemental - values like subordinate and riffling Motional-value, temperature-value, force-value, pressure-value, and so on degree-wise section and sub-sectioning to the **infinite effects**, efficiently qualified to accord their efforts for the preparation and development of this Universe.

{(*Here, concerning the sense of* ***'infinite effect,'*** *it remembers a technical point from the book series that any imaginable or unimaginable sense as a most micro level of 'moment' leads to derive some form of physics* (that is, energy value or elemental value) *due to the most fleeting roles of energy elements whether it sensibly accept as effectual or ineffectual.*)}

i. The initial phases' performance of 'Nullifying force of energy' is one of the pre-eminent features and actions, which is naturally assessable in the developed stages stance as force elements or elementary levels 'working essence' that perform the functionality of the sense 'force/s.'

(The objectual word sense nullification means invalidation of the value of something, depriving its effectiveness, or continuing its existence as appearance.)

→Here, from the standpoint of disaffiliated or affected units of Pure-energies, the **'Nullifying force'** may be defined by certain types of very primary phases, 'compulsive-energy-signal' derived while enrolling differential values functionalities due to footing or formation of energy elements of their diverse dispositions.

The **Nullifying - force** tends to resist breaking and care to entail cohesion to recoup an earlier alikeness status as Pure - energy.

In the initial stages, 'Nullifying force' is performed as one of the pre-eminent or leading acts, which is also consistently active in the matured 2nd-energy-belt known as the developing phase 'force - elements' or types of working essences that perform the function of word sense 'force/s.'

→Gradually, with the accumulation of lymphs and stepwise growth of lymph-energy bubbles, they could attain certain resistive impulsion to deter the effect of 'nullifying force' from afflicting energies.

→In the meantime, due to the internal cohesiveness of the cohesion of **Lymphs** and Lymph-energy bubbles, commence to adhere to other differential values of lymph and lymphatic energy bubbles. *("Please note that the word* ***'lymph'*** *and its related terms will talk in the next.").*

→By the by, here to cite in brief that internal cohesiveness accomplishes due to the footprint of generic values' that is, the amount of functioning generic - energy bestowed from Pure - energy.

ii. 'Bullwhip effect' and formation of prosperous /advanced stages elements.

("Some analytical clarification."
The term "bullwhip effect" typically refers to a well-known supply chain phenomenon. This phenomenon involves the amplification of demand variability as it moves up through distribution channels from retailers to manufacturers and raw material suppliers.

"Also, the same sense described based on classical physics, when a person creates a wave in a whip by sending impulses, the loop or the wave moves along the length of the whip, gaining speed due to a mini-sonic boom created by the supersonic motion of the end of the whip.

Here, the kinetic energy of the whip converts into sound energy, and as the whip breaks the sound barrier, the energy disseminates to the surrounding air.")

i. However, from the perspective of baby stages of the Universe, the sense 'Bullwhip-effect' is utilized since there are incremental swings of effective forms of "energy - elements" and their elemental - values throughout 'IEfu' or influencing energy areas of the Supreme - Universe, in consequences of the diversified act of afflicting - energies, lymph, lymph-energy-bubble, lymph-clump, retentive unique-items, and many other supplemented-items, which all were once at Pure-status that is 'Absoluteness-status.'

As a result of the consistent process of cohesion, adhesion, disintegration, coercion, compression, and external surface tension from afflicting - energies and some freelance lymph allow the lymph-energy-bubbles to turn into Lymph-clump. Besides, they keep engendering various energy - elements incrementally that cause the initiation of the "Bullwhip effect" inside the spatial of influencing energy (IEfu) in the Supreme - Universe, which is the zone of IEsu.

ii. Here to note, during the advancement process and in the procession of increasing differential - values criteria. Some energy - elements and elemental - values of units of energy gradually improved their effectuality than the primary stages of infantile energies of the fecund-Universe.

→Further, consistency of asymmetrical /irregular activities within the assortments of afflicting - energies, lymphs, lymph-bubbles, lymph-clump, elements, and so on derivatives initiated **churning processes**, consequentially emerging of variable elemental - values of energy elements like temperature-element, pressure-element, and so on to many.

*(Here, the word sense **'churning process'** is standardly known as shaking milk to change and separate butter from*

milk. But technically, the churning process isolates the lighter particles of a suspended solid from a liquid like milk.)

All those elementary levels of energy elements cause to initiate gradual changes to all forms of energies, to lymph-clump, lymphatic mass, and thereon, stepwise further course of actions, the appearance of mass-forming-energy and pavement for, or accumulation of break-off and new genesis in the form of masses of energies within encompassing areas of the baby Universe.

6. Initiation of constructing Gestation line.

Besides the roles of **churning** activities, simultaneously due to the engagement of the **wheeling effect**, some non-effectual and non-participant energies are very close properties or almost Absoluteness-status of Pure-energies, like extremely short swaying unique items, lymphs, pro-tem /stopgap units of energy and likewise many close to very close units of 'Pure status' mean **ineffectual or indolence** energies on the term of physics, energy-value or elemental value, begin to appear.

While in multi-directional physics values or energy values roles throughout, those **ineffectual** or **indolence progenies** tended to drive away from a place of origin to reach the juncture between the IEfu and the Supreme - Universe. Also, many of those **ineffectual** or **indolence progenies** in some way succeeded in advancing and docked between the 'Pure-energies' of the Supreme - Universe *(Esup)* and areas of 'Influencing-energies' in the Supreme - Universe or (*IEfu*) and bit by bit, those relatively Pure-status energies (*that is, ineffectual or indolence progenies and likewise many other*) managed to represent as contributors to form the Gestation-line or surface barrier of the embryonic Universe *(somewhat attempted to indicate in the drawing SU20).*

*[{("**Some analytical clarification.**"*
Based on the objectual sense, the ***'wheeling effect'*** *means asymmetric thrust in a rotating manner. But, from the viewpoint of energy, the word* ***'wheeling effect'*** *is somewhat similar to repetitions or in the rotating manner of activities that disturb the originalness of energies. Also, it allows for* ***reorganization, from near to very near originalness*** *of the belongings throughout the progressing Universe.)*

{Here, briefly, it may quote the point ***'reorganization, from near to very near originalness'*** *does not mean 'entirely equal' that mainly applies following the practicality of* ***objectual*** *(here,* ***objectual****, basically pragmatic on three-dimensional purview) sense since on the standpoint of affected Pure status of units of Pure-energies from Absoluteness-status, its componential (mean, energy elements) disturbance gradually broadened to an* ***uncompromising reversal*** *state* (here, **uncompromising reversal** means 'multidimensional perspective'*) increasing the objectual sense as infinite diverseness.}]*

i. Effect of differential value and the tendency to equilibrium along with the increasing motion, or say, growing 'Motional values' of motion-element of energy derived with appearance, disappearance, and reappearance of many known and mysterious energy-elements & effective outcome recurrently.

Act of those **elemental values** and its effective outcome of energy elements may represent minimally this earthly available lowest possible measuring unit like the satisfying word-sense yocto, Planck, or beyond as unimaginably tiny unit to biologically evolved intelligent species in this habitable globe.

Those typically smallest **elemental values** (*that is, somewhat analogous to the objectually recognizing most* *micro levels of physics values, which are standardly derived when objects are in action)* gradually acted on "lymph and lymph-energy-bubbles" with the introduction of lymph-clump, or say, the initiation of **tuffet lymph-energy,** which is a stage before the formation of mass-energy in the influencing energy (*IEfu*) of the Supreme - Universe.

→For the convenience of comprehension, it may hint at the above or present in an overall or summarizing way regarding the **stepwise** changes and the procession of embryonic stages energies in the Influencing energy zone of the Supreme Universe (IEsu), somewhat like: -

►Interaction of Symmetric and asymmetric motion initiated to disturb the State-of-Absoluteness in spatial of the Supreme - Universe.

►The formation of **unique energy levels' products/items** and incrementally changing afflicting - energies allow for advancements of energy values or elemental values of energy elements or physics values incessant functionalities, all those assisted in structuring the Universe from the embryonic stage of the Supreme - Universe.

Some of the **unique energy levels' products/items** are like: -

►Lymph.
►Lymph-energy-bubble /Lymph-bubble.
►Lymph-clump.
►Tuffet lymph-energy,
►Lymphatic-mass.

►Likewise, many other energy progenies or energy levels products retain from shortest to long-lasting upsurges in the 'IEfu,' which all partook for structuring the Universe from the preparation of the Gestation line then gradually turned to its mature stage as **individuality**, from the innately featuring state-of-Absoluteness of the Supreme - Universe.

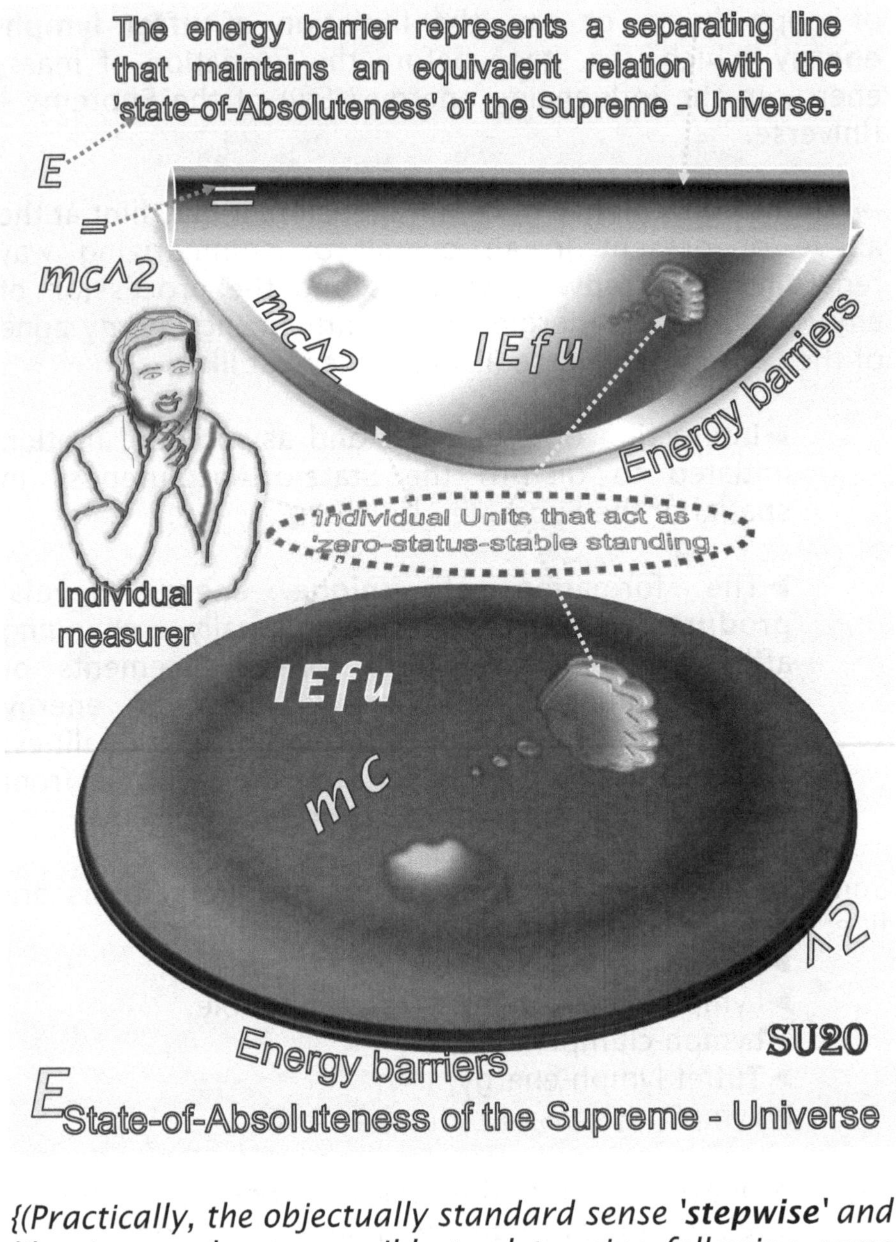

{(Practically, the objectually standard sense ***'stepwise'*** *and likewise words are possible to determine following some measurable method of counting the time. So, it will be incomparable and incomprehensible if someone attempts*

to apply the same and many other objectual senses to the vast areas of 'energy-Ocean' or the multidimensionally functioning energies of the IEfu.)

*(And, from the comparable assessment, the expression of **individuality,** that is, 'distinctiveness' technically on the aspects of the Supreme - Universe, the overall primary stages energies of the IEfu, and many other individual units to the standard sense of living beings who are the acting analysts in this habitable nature may work as almost similar.*

Thus, on the aspects of parametric implications of individuality, all the above individual positions as the Supreme - Universe, the IEfu, units, and the living beings represent somewhat like 'zero-status-stable value' for a consensus measurement and satisfaction since all the above three or more items /values act as zero or null position whatsoever the items integrally or constitutionally symbolize as unimaginably diverse, absolute zero, or infinite standing like the state-of-Absoluteness of the Supreme - Universe.

The drawing SU20 will give a comprehensive clarification on the above issue.)}

ii. "During the process of gestation growth, some lymphs and many of their progenies find ways to reach the closest surface of the Pure energies of the Supreme Universe. This gradual improvement leads to the distinctiveness of the 'Influencing-energy (IEfu) of the IEsu zone,' forming and aggrandizing the gestation line, as indicated somewhat in the drawing SU21."

→Further, while in the course of slowest development and gradual procession toward forming the Gestation - line. Those surface areas' lymphatic mass turns into readymade items as mass-forming-energy and mass-energy, or these masses may mean /sense as a form of collective unit/s of puerile, adolescent, or somewhat the near Pure-status energies of the preparatory or embryonic - Universe.

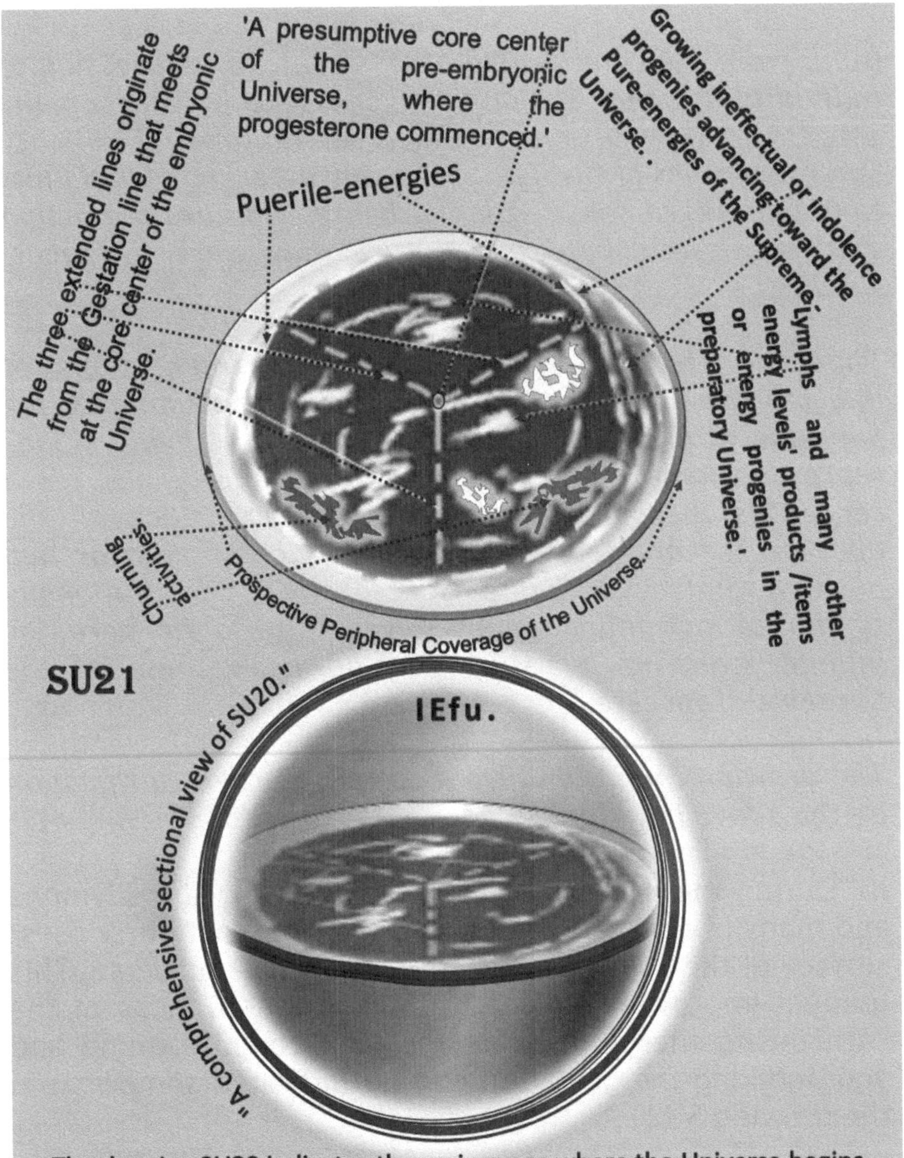

The drawing SU20 indicates the main areas where the Universe begins to form in the 'state-of-Absoluteness' of the Supreme – Universe.

(Here, one point to recite is that multidimensionally functioning elemental values, energy values, or conveniently assumed as objectually standard physics

values roles, contributed to constructing the Universe from the initiation of gestation in the womb of the mother Supreme - Universe.)

→Depicting with a drawing SU21, it attempted to indicate the embryonic - Universe or IEfu, where due to some primary stages components and its growing activities make possible the advancement of the prospective Universe.

In the drawing, the three extended lines originate from the gestation lines, which meet at the **core center** of the embryonic Universe. Those lines are displayed mainly to indicate following the trio-dimensional or 3D conception of specifically evolved intelligent species in this biological globe.

*(This **core center** is the presumptive position or spot where the gestation initiates in the prospective Universe.)*

It also may mean that a three-dimensional conception of living beings works as the practicality for fair acceptance, which predominantly governs the observable items based on the ways they evolved in a place of liveable habitat.

iii. "Some more notes concerning the preparation and formation of Gestation-Line, all around the prospective region of Influencing energy or 'IEfu' in the IEsu zone of the Supreme - Universe."

At the most primary stages of the Universe, the restive and sluggish lymphatic units are least differentiable due to the weirdness of meager acts.

The least differentiability also resulted in slow progression and weakest instigation due to irregular energy values momentary efficacy comparing the centralized or the epicenter of 'Influencing-energy' of the gestating-Universe or IEfu.

It also may mean that the formation of the Universe commenced at a particular acute center point. And thereon, the unimaginably slowest process approaches **infectiously** to unimaginably expanded areas *(here, **infectiously** means 'something like the invasion and multiplication of diversity' by disturbing the Pure or originalness and the units of afflicting Pure energies within the IEsu zone).*

→The above energy levels' technical clarification is clear and understandable to the units of living frames represented as bio-intelligent species that (mean, *graspable aspects*) have evolved into alive systems due to the blessing of the typicality of essences or resources available in this bio-adaptable globe.

As a result, the commonly accepted practices that illustrate the logical and apparent meaning of words, such as **'expansion, infection**,' and likewise senses, could assess comfortably following the principalities of observable items and analyze within some celestial circumscription.

Also, considering the energy, all those perceivable word senses like expansion, contraction, infection, diversity, and pure or untouched features outrightly concluded or terminated upon maturity of the Universe within the specific coverage due to the **'no begin no end' state-of-Absoluteness** of the mother Supreme - Universe.

*((Here, objectually suggested '**timelessness**' may somewhat denote similar to the attribution of 'state-of-Absoluteness,' where only the motion element plays its perpetual neutral roles to maintain the function of sense True Soul of other oneness of all 'energy elements' that is the totality or absoluteness of the mother Supreme – Universe.*

*A little more technical clarity about the philosophical sense of '**timelessness**' and say, at energy levels, Time is eternally integrated with motion elements, whether non-countable for the standalone position of motion elements in the state-*

of-Absoluteness of the Supreme - Universe or countable in the distinguishing manner where there is the issue of diversity like the matured Universe, and specifically in the 2nd-energy-belt.)}

iv. Based on the standard natural borne ideology, it may speak that some of those lymphatic units formed exceedingly close to the symmetric area of the Supreme - Universe, meaning next to the asymmetric peripheral line of the preparatory embryonic -Universe before evolving as fecund-Universe or extremely slow-growing 'IEfu.'

Those Lymphatic units tend to push away or assume to gravitate away because of the initiating **eddy effect**. Or the same may sense as **a counter-flow** effect from the active regions and attempt to fuse with the stagnant /static Pure-energies' state of Absoluteness' of the Supreme - Universe.

However, ineptitude acts due to the near completeness of Purity to those restive lymphatic units and many other remnants of indolence unique items gradually lined up at the certain inerrant edge of a null-effect Pure-energy zone or the perimeter of IEsu.

*[{("**Some analytical clarification.**"*
By the way, here, technically, it may clarify a little and say, on the point of energy levels Physics, state-of-Absoluteness proprietorially represents a physics-less status or something recognizable as (abs0) (or like objectual absolute-zero), which is possible to depict by the symbol E (in all instances, 'E' considered to be referenceable as a frame like the Supreme - Universe) that maintained equivalent or parallel relation to tackle the irrevocable diversities in the coverage of IEfu (presume as ^2) among the containment indicated by mass (m) and motion (c) that follow the excellence of equation E = mc^2.)}

{Also, qualitywise, one significant point to quote is that the formation of the energy level's mass (m) and motion

element (c) at the gestating period of the embryonic Universe is quite unlike the masses (m) and motion arousing essences (c) of developed Universe, especially in the matured 2nd-energy-belt, where the objects or objectual items regardless of most micro to a massive structure that represent from short to a long term, simple or compounded frames of mass-energy.

Thus, most of the fundamental stages of energies and working essences or energy elements gradually lost their Purity and originalness, aggravating the quality and leading diversity, which is apparent in the matured 2nd-energy-belt.}]

Nevertheless, briefly say, consequences of the utmost slow evolution and devaluation during the gestation process, those generated lymphs, lymph products, and unique items technically performed to construct the Gestation-line that allowed structure-up steadily from the embryonic period to the maturity of the actual Universe.

v. Initiation of crest /turf and Gestation - line.

Laconically say, since there are unimaginably prolonged commotions among the inequitable units, some specific types of shake-off /escaped energy - values (*or objectually applied some of the infinite types of physics values*) from afflicting energies in the pre-embryonic Universe slowly and progressively partook to form **crest /turf** for initiation of the early stages' **energy - barrier** between the **symmetric** and **asymmetric** areas. Or say the same, that is, across the space separating **symmetric** areas of the Supreme - Universe (Su) and **asymmetric** region of the growing stages embryonic Universe or justify it as 'IEsu.'

This **energy - barrier** may also be considered the additive of the initial stages' **gestation line** of the embryonic Universe, which gradually developed and fixed at the extreme edges of the 'IEfu,' closest to the symmetric areas of the Supreme - Universe.

(Technically, from the objectual viewpoint, this primary stage's gestation line or lately framed energy barrier acts as a protective fender between the two distinctive uniqueness of 'Su' and 'IEfu.')

Due to the very long period of constructive role from the preparation and towards maturing of the tender ages' Universe, this gestation line could able to nurture with the balance due to multi-dimensional energy values *(or objectually considered Physics values)* applications through the edges of the Supreme - Universe to maintain almost equitable features on rhythmic patterns, for the sustenance of internal compositions and their effective or ineffective numerous roles throughout the preparatory Universe.

*[{("**Some analytical clarification.**"*
Further, it is to quote that the above differential functionalities within the womb of the mother Supreme - Universe or 'IEsu' zone are attainable to apprehend following the profound meaning of the most famous equation of twentieth-century physics that is, E = mc^2, which can justify in little advanced in-depth of energy levels, where 'E' may present as the Supreme - Universe since proprietorially it is the 'state-of-Absoluteness' with Pure-energies that serve the eternal position with the ideology of 'no begin no end' assignment, value, infinite and likewise many similar to the sense as absoluteness quality and quantity.

*This 'E' maintained equivalent (=) relation whatsoever the energy levels upheaval due to growing diversity to the distinctiveness of masses (m) and the dynamical activities for the motion deriving energy element or motion element (c), within the entire coverage (**^2**) of the 'fecund-Universe' or IEfu.*

*{Here, the sign ^**(lambda)** standardly is designated with a figurative area denoted by '2,' or this numeric value '2,*

maybe supplemented or dissuade (using the sign >, or <) *as appropriate to comprehend following the advancement of intellectuality of living species in different habitable natures since the sense intellectuality is simply a mathematically apprehending variable constancy, which keeps on changing following the deviations of standardness of atmospheric conditionalities or the parametrically determinable criteria that confine the meaning of intelligentsia.*

Simple, like a naturally evolved full-bodied wise living being, their discovered Artificial intelligence (AI) and other classes of intelligent units all will begin to deviate or lose their standard potency immediately upon crossing the parametrically determinable ***atmospheric*** *limitation/s* (here, the **atmosphere** is something the micro momentarily active surrounding elemental-values, energy-values or physics values essences like temperature, pressure, resistance, etc∞., which in reality at energy levels, momentarily appears and alter their strategy following the conditionalities of surrounding atmospheric physics (*means, Twig + Energy atmospheric physics).}]*

vi. The Gestation line gradually turns murky /obscure.
The foremost primary stages' energy barrier, which has taken the pridefulness of the Gestation line, gradually turns into a **murky** station.

This "**murkiness**" or obscureness of the Gestation - line is due to the proportionate increase of diverse **releases** during unease activities among the disturbed energies throughout 'IEsu.'

Also to mention that throughout the dynamic areas of the IEfu, many of the energy levels' **vital releases** could somehow be **flown** or assumed as **sliding** manners, managed to reach the peripheral zone of the Universe that consequentially participated in the progression and construction, which progressively embedded and gradually allowed to achieve in terms of strength of the Gestation - line and successively became full-fledged a robust energy barrier for the Universe.

*{(Now, the quoted word sense '**vital releases**' indicates that some of the dispersal most fundamental 'energy essences or energy elements participated in reorganizing or managing for curations of the peripherally situated most minimal affecting or afflicting units of Pure state energies to reach the closest features of Absoluteness-status as Pure-energies of the Supreme - Universe, and thus the internally deriving divergence roles of the IEfu gets hindered from impacting the actual symmetric areas of the Supreme - Universe.)*

*(Here, one thing to requote the objectually standard word sense '**flown**' and '**sliding**,' and likewise, typically thought to function in vacuum or empty like constancies or invariable values. Those objectual standards are practically far from the reality of energy level works, where some form, on the point of the parametrically distinctive as lowest to the highest degree on the aspects of **quality** of energies continuously are in operation.*

*(Also, the objectually standard sense '**quality**' conveniently may justify based on the different phases or degree-wise as dense and light status, which functionally or on the point of objectually applying sense 'physics' may be effective or ineffective to some limits of bias /unfair or unbiased /fair requirement to intelligent species.)}*

Further, at energy levels, continually due to the sequential activities with multifarious energy values or in-depth **elemental values'** releases, were able to arrive and coasted in the peripheral zone, which resourcefully or smoothly somehow could participate or manage to alleviate or reorganize afflicting Pure-energy units that proprietorially allowed them to turn up as nearer to the 'Absoluteness-status' or its equivalent as the units of Pure-energies and added with the energy barrier that also constructively became a robust challenger for internal uproar in the 'IEfu.'

Technically, the progressed Gestation line, energy barrier, or that justified as a robust energy fender, analogously a high-octane than the low-octane internal containing in the growing Universe. Since, on the aspects of physics or physics-values, energy value, elemental values, or simple way speak as formidable work potent, internal containing of advancing Universe, gradually loses their significant energy elements or working essences that consequentially drove away towards the periphery, compelling the internal energies to turn into low-octane dispositions.

(Regarding clarity on energy level's working essences like 'energy elements and elemental values,' elucidated in the earlier book series titled 'Energy and The Complete Universe' and will be discussed more, as and when appropriate.)

So, briefly say, the Gestation line occupies distinctive murkiness /obscureness due to the insignificance of differentiability with the symmetric line of the Supreme-Universe.

Also, conveniently and for ease of comprehension to bio-intelligent species, this naturally borne word murkiness or may be justified as murkiness-zone is applied to individualize the visual sense of living beings from the actual Pure-energies of the Supreme- Universe and the distinction of perturbed energies of the Universe.

→In the IEfu, various energy levels released and consequential outcomes gradually accelerated due to diverse energy values functioning *(or the same sense may apply as objectually realizable or unrealizable different forms of physics and their deriving products),* emerging diffusion, fusion, jumbled motioning effect, and many other exercises through repetitive diluting or concentrating processes during the multiple energy levels activities.

Also, in the IEfu, besides the emergence of many energy level outcomes while in primary stages' rumpus activities, there are inferences of many other perturbed and unperturbed units of energies, which gradually pushed through the afflicted pure-status energies and consequentially some of those added successively or in-line to fortify the Gestation line or energy barrier of the Universe.

So, concisely quote that the whole areas of IEsu slowly and rhythmically turned to a turbulence zone since there is incremental **splitting or breaking and lessening** of the originalness from the Pure status.

Or the same point concerning **splitting or breaking and lessening** of the originalness is possible to define as the incongruency or inappropriateness of the elementary level of energy elements from the earlier stages justifiable as the units of Pure status energies, which may presume as an eternal constant that suits to specifically structured intelligent species in different habitable natures.

(For clarity, it is to quote that the word-sense 'Pure status' is a somewhat distorted state and not the actual 'Pure-energy,' which is available only in the Supreme - Universe.)

(Here, analogously, one thing may note that the objectual word 'volume' or on the point of some determinable dimension/s, or the sense as an objectually ascertainable amount of 'space' a substance occupancy, is reasonable to analyze and understand, maintaining the sense blank/ expressionless /vacant as a constant for nonempty energy levels roles in the eternally active matured 2nd-energy-belt.)

Also, it is to speak, from the perspective of the energy, practically everything, even the term extremely tiny level's momentum, performs sequentially following the multi-directional functionalities maintaining the state-of-Absoluteness of the Supreme - Universe as the fundamental base for the immature activities of the Universe.

→However, living species in various natures and natural ways of their evolving sensing faculties and psychological roles tend to secure any finding /conclusion, following the analyzation and measuring approaches within the availability of the inbuilt quality of living grades of energy and the direct or indirect, overall represent as inductive quality in reliance on some justifiable or **dependable base**/s.

Or the same **dependable base** may justify as conditioning **zero-status-stableness** or something like the predisposing-constancy position, whether it (*that is, **zero-status-stableness)*** represents as own-self the living unit, measurer or target, within accessible surrounding directly, and indirectly even to the remote area's imaginable or unimaginable entity.

vii. 'Note in brief about the word-sense dilution and its constructive roles brought massive changes to facilitate energy levels numerous applications, like the creation and functioning of Influencing-energy and its in-depth or the most fundamental phases' working essences, that is, the Influencing-energy-elements.'

The objectually applied word sense '**dilution**' is standardly acknowledged as an action that makes something weaker on the aspects of functionalities from the earlier standard or strength.

For the Universe, conveniently, the same expression '**dilution**' is applied to convince the energy levels attribution, where energies gradually weaken in terms of resiliency from their original status where all working essences or energy elements are at their **decisively unified Pure status** (***decisively unified Pure status*** *is a deviated form, and not an actual or truly Pure-energy).*

At the primary stage of the baby- Universe, **dilution** is due to the deprivation of one or more 'energy elements' from the originalness or something like the pureness of any assumptive amount or unit-value energy.

Or it may be considered as the loss or **dilution** of ‘energy elements’ due to the non-performing, weak-performing, or sequentially weakening /declination of elemental -values from a state of originalness.

Or simple way may speak as the adequate losses of energy-value from a unit or unit-value energies, leading to drive away from their originalness or Pure status, which in turn, slowly and consequentially administered to transform or sublimate into a new status as 'Influencing-energy' or likewise, in the fecund-UnIverse that is IEsu.

Likewise, the same expression '**dilution**' may consider some form of loss from the unit/s of shivering, pseudo and bonafide (if) pure form of energy in the "IEsu" or influencing energy zone of the Supreme - Universe.

Even in the developed Universe, the word-sense **dilution** may be considered or incorporated as elemental-value loss during various activities of units of energies at their divergent status.

Similar to the word sense '**dilution**,' there are many linguistic expressions available to the modern intelligent generation that are appropriate for utilization to distinguish and understand the work and performance of diversifying phases of energies.

*[{("**Some analytical clarification.**"*

*For conceptual clarity, it can speak that within any form of unit-value energy, the motion element represents the lowest level of any scaling degree, whether determinable or indeterminable. It (*motion element*) is the most fundamental or* ***credential*** *working component, closely followed by the ascendant influencing energy elements of the influencing energy.*

It means that if someone thinks to quantify unit-value energy from the bulk state energies of the Universe specifically from the 2nd-energy-belt, then there must be a componential minimum of ***two elemental pieces and placement*** *in that presumptive unit like the most fundamental and* ***credential*** *one is the motion element, and the level-wise 2nd closest one will be the influence-energy-element of influencing-energy, as indicated in the drawing SU21a.*

(In this regard, someone may clear the above ideas concerning the componential minimum of ***two elemental pieces and placement*** *from the three-dimensionally projected sketches SU14 and SU21a and will define more as and when appropriate for placement.*

Indeed, if an individual attempts to realize the drawings from a multi-dimensional perspective, then the 3D standard intellectual conception has to compromise and abridge to some level of satisfaction to comprehend the most complicated energy levels' momentary works.)

(*Here, the expression '****credential****' is utilized on the aspects of the working role to indicate the persistent activities, comparing the innumerable less credence energy elements that relentlessly function as versatile activities.*)

{Also, to quote, based on the standard 3D notion, someone of the present day's intellect species may think of quantifying the motion element to its solo state but, in short, may speak of it as an impossible task in a practical

sense since the status of energy is omnipresent to its myriad form in the matured 2nd-energy-belt.

The motion element is not a static or free-flowing entity but is consistently vacillating with other infinite numbers of varied types of energy elements in the vast ocean of energy.

Or the same may simplify and speak based on the standard notion of living beings in this livable nature of matured 2nd-energy-belt. Whenever someone thinks and expresses 'energy,' it logically or generally represents some discrete amount as packets, sacks, units of mass form energy or mass-energy, which may or may not regarded as a referable frame like object/s.

However, energy elements are more the deep-down working essence of those mentioned unit values packets, sacks, units as mass-form energy or mass-energy frames referable like object/s and represent some acknowledging 'state of matter' by intelligent beings of this biologically habitable nature.

So, from the perspective of energies in the vast areas of the matured 2nd-energy-belt, productive natures comfortably could find access and quantify energies under the conduciveness of atmospheric conditionalities and ***integrally*** *organize the amount into unit/s as mass form energy, mass-energy frames, and later on detectable and undetectable objects of different shapes and sizes.*

*(Here, the expression 'i****ntegrally****' is considered on the point of energy elements or working essence, including 'motion element.')}*

("The indication of logical expression 'causalities' for boosting diversity from the initial stage of founding the Universe.": -

For the clearance of conceptional hitch, it may indicate that the 3rd-energy-belt occupied only one type of working essence or energy element, the utmost finest form of energy and functionally dignified by motion-element. And, the IEfu or Influencing-energy of fecund-Universe mainly loomed by the Influencing energies.

Hence, on the aspects of energy and proprietorial meaningfulness of expressions 'motion' and 'influence,' both of their elementary levels' qualified to engender more and more diversity, exercising shuffle and reshuffle of the containing even from the embryonic stages of the Universe.)

(Basically, the drawing SU21a figured following some discernible dimensions, but multi-directionally, the influencing energy element can move in any direction, ensuring by the external energy levels applications within the small or large limiting area/s.

Concerning multi-direction and multi-dimension, I have clarified in different ways in the earlier book series titled 'Energy and The Complete Universe.')

(Indeed, based on the typicality and constructional aspects of intelligent species and bio-living beings in the matured 2nd-energy-belt, it may speak, at energy levels' 'energy elements' are multitasking considering the point of characterization and impact that follow the diversifying elemental or physics values numerousness roles.)

*{Besides, one thing to quote is that the 'motion element' is the only one that is 'eternal and self-sufficient' in performing impartial administrative roles wherever the creation of any energy levels elemental (*that is, energy elements) *congregation.}*

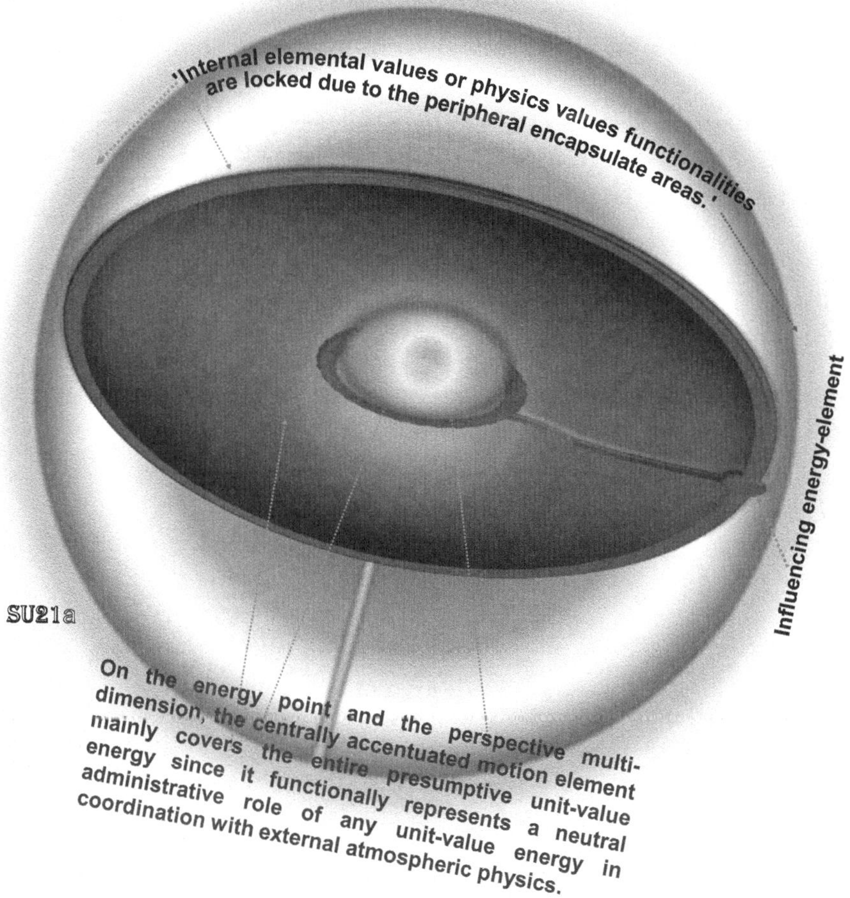

(Also, briefly, it may hint and say that whenever some amount of energies tend /manages to encapsulate as a unit form, instantaneously, the internal physics of the unit-value energy gives life by organizing the energy elements or working component, including the motion element in concurrence with the conduciveness of surrounding atmospheric physics.

Thereon, the journey begins of the encapsulated unit-value energy towards progression following favorable circumstances of atmospheric physics to become mass-energy, mass-energy frames, and then gradually and consequentially of multi-levels implementation of energy physics that lead to transforming into objects of different shapes and sizes.)}]

viii. Preliminary-surface barrier or Primary-gestation-line gradually transformed to serve as an eternal constancy for the actual Universe *(drawing SU22)*.

Energies of the primary gestation line, **energy line**, or preliminary-surface barrier act as consonant or reconcilable states between the mother Supreme - Universe and the embryonic Universe. This "**energy line**" may also be called the **rudimentary Pure-energy coating** between the 'Supreme - Universe' and the 'fecund Universe' indicated in the drawing SU22.

Or this **rudimentary Pure-energy coating** may be designated as a comprehensively near **absolute-pure-energy zone** that is **Eabpu** of the IEfu or progressing embryonic - Universe, which (***rudimentary Pure-energy coating***) remotely commenced in the IEsu or prospective Universe during incubation, or upon actual disturbance or infection begins at an acute point due to slow depositions of disturbed symmetric function in the growing process, which slowly and steadily later stages' reached at a consolidated position to turn into near-absoluteness-status or the alikeness feature of the Supreme - Universe since there is the harmonized result of internal energy levels physics of **Eabpu**.

*[{("**Some analytical clarification.**"*
*Here, **'Eabpu'** is considered the near absolute-pure-energy zone /layer /coating of the primary stages of the Universe.*

*Technically, from a broader perspective and for the progesterone period, it (**Eabpu**) can be assumed as a conceptional or ideologically presentable section /unit /part, differing from the true purity of the Supreme - Universe. This **'Eabpu'** is a zone of 'acting pure energy' proprietorially almost /nearly absoluteness to the Pure-energies of the Supreme - Universe, or the same may dignify as a **rudimentary pure energy layer** between the Supreme - Universe and the Universe.*

Besides, on the aspects of energy levels physics or energy-Physics, it is to quote, due to the adequateness of **own** (here **own** mean, Eabpu zone) *compelling physics comparing the growing fecund-Universe and its engendering physics, the 'Eabpu' zone is exceptionally sturdy to challenge and check any form of functional roles due to internal diversity of the IEfu or baby stage's Universe.)}*

{(By the bye, technically, from the perspective of expression 'physics,' the Supreme - Universe may be dignified in the short title as ***'Oabsp'*** *the 'state-of-Absoluteness' or true* **Symmetric** *(or following the objectual sense, the expression* ***'Symmetric'*** *can be simplified as proportionate and balanced), and bearing ever-constancy in the term of both numeric and magnitude.*

(Indeed, the original property of the 'motion element' is incessantly neutral, regardless of whether it pertains to intelligent species at the incredibly lowest or highest levels of ascertainable or beyond some degree of dynamism, so sensibly it does not affect the ***symmetric*** *undertaking of the endless station of Supreme-Universe.)*

The denoted value ***"Oabsp"*** *for the Supreme - Universe may be defined, assuming the ideologically represented number symbol (0) is at an 'absolute Pure position' and it enlightens that the Supreme - Universe itself is at an absolute-equilibrium state, or may speak, like an unconditional Absoluteness-status, or also can signify as a state of 'Oabsp.'*

Or on the aspect of 'physics,' the ***'Oabsp'*** *represents the 'supreme or absolute' immutable position for all forms of 'physics values' impact that may emerge inside the progressing and (if) anything from the actual matured Universe.)}]*

Also, the units of energies of the extreme exterior of the 'fecund-Universe' or the preliminary-surface-barrier, primary-gestation-line, or 'Eabpu' are least affected or incongruent and have immiscible features. It is because the unit of energies in the areas of the surface barrier has lost extremely minimal criteria from the originalness of Pure - status.

So, the **inefficient** units of energy in the preliminary surface layer initially and comprehensively remain much **denser** than the internal holdings due to increasing gestational activities in the embryonic - Universe. But, with the growing process, those unit-value energies of the preliminary surface layer gradually balanced up and ultimately turned into normalcy to near-absoluteness-status, which is entirely negligible or relatively meaningless in the aspects of characterization and impact and similar to the 0absp.

*{(Here, for energies of the preliminary surface layer, the objectual word-sense '**inefficient** or considered as counterproductive behavior' indicates a sturdy station on the point of physics values act and effect.)*

*(Also, at the energy levels, the expression **denser** is entirely unlike the observable sense regarded as closely compacted, thick, and undiluted. Or the same (*that is **denser**) *at energy levels may somewhat resemble and express as entirely an impenetrable zone /place for all sorts of external physics.*

The other ways the word sense, thinner, uncompressed, and so on, which at energy levels may justified by the expression as reasonable, easygoing, tempered, or intense on the aspects of physics values activities.)}

*(One puzzling but significant point to note is that at the energy levels, the actuality of the term '**physics,**' which appears momentarily following multidirectional passages, maintains conditionality for generating diversities during the emergence of multiplex energy values or energy physics. Therefore, for objects and objectual products, or based on the standard observable sense, the intellectually presented expression '**physics**' culminates in satisfactory conclusions.*

For the above, as a sample drawing SU23, depicted to comprehend how complex the actuality of the term 'physics,' which works at elementary levels of energy elements, in the depth of an object or mass-energy frame upon forming and functioning, maintaining remote /indirect relations with the vast coverage of energy -Ocean. However, an observer or analyzer standardly acts as a center (0D) or 0-dimension for an overall value by studying the limited right and left (1D + 1D) of direct accessibility.

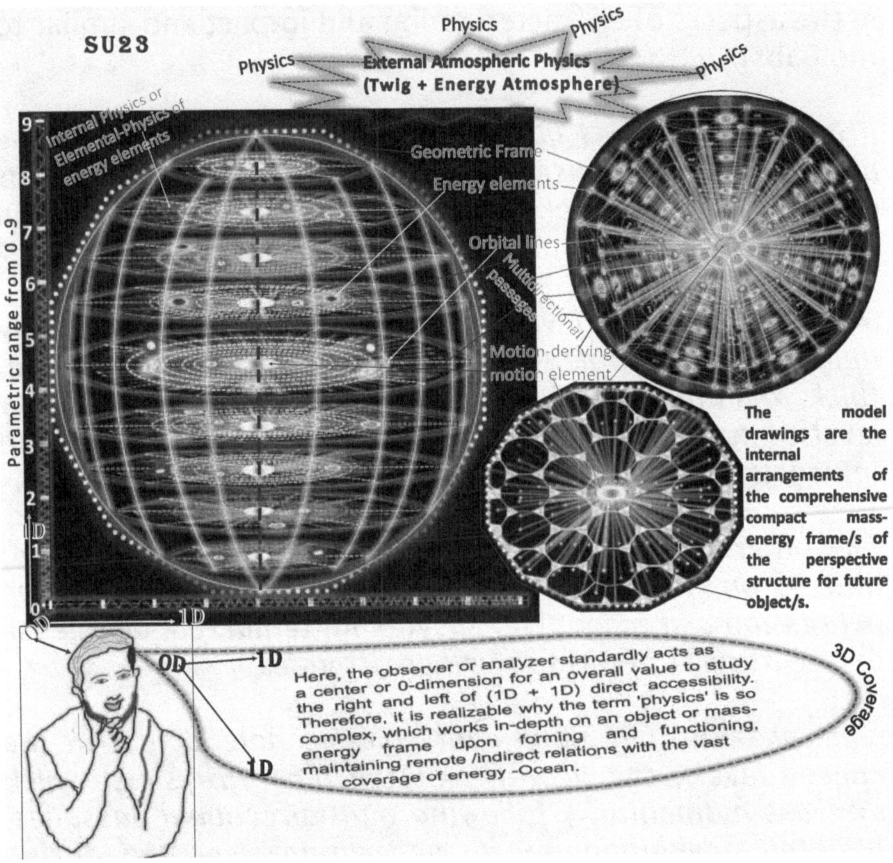

ix. The gestational activities help to organize the structural foundation of the prospective Universe by forming the preliminary stages' surface layer, which subsequently develops into a matured and sturdy energy barrier or gestation line.

Additionally, gestational activities lead to a considerable loss or dilution of the deep-down of most essential and unstable elements and their intrinsical elemental values of near-pure status energies and other units of Pure-energies within the IEsu. It means that from the commencement of gestational activities in a slow and steady process, all the contained energies in the IEfu, regardless of their fully, partly, or non-affected units of pure energies, become dislodged from their original state.

→At energy levels, **consistency with synergisms** of differential-values activities and the endeavor for reconciliation to obtain a status of equilibrium among internal holdings in the embryonic Universe compel a recurrent process or keep re-processing the internal disarray so the containing can readily justify as energy elements.

*(Here, "**consistency with synergisms**" suggests a stable, reliable state where collaborative interactions continuously enhance overall outcomes.)*

Mathematically, this concept **("consistency with synergisms")** may be expressible using a polynomial function, in addition to polymeric and non-polymeric conditioning among the differential-value units of energies throughout the 'fecund-Universe.'

7. The significance of the reasoning expression 'equilibrium' is genetically or inherently obtained from the Pure-energies of the Supreme - Universe.

"The activity and attempt to reconcile 'differential values' and achieve equilibrium status serve as the foundation for validating all conclusions in modern juvenile science."

Alternatively, for intellectual species in any livable nature of the matured 2nd-energy-belt, it may be simplified as follows:

At energy levels, there's rarely or no possibility of achieving a pinpoint accurate conclusion for anything. Instead, conclusions predominantly rely on overall parametric understanding. At energy levels, **everything** functions consistently, whether discernible judgments recognize it as reasonable or deny acceptance based on the limitations of direct (*or* ***close circuit***) and indirect *(mean,* ***remote circuit****)* ways of involving physical realities.

*{("****Some analytical clarification.****"*
Here, the expression ***'close circuit'*** *may be exemplified with the regularly attainable practical items and events through the physical system, and* ***'remote circuit'*** *is far off from direct observation but ideological certitude, which happens causally for some inexplicable reasons.)*

*(Here, the expression "****everything****" is not limited to tangible outcomes of "physical physics," which are comfortably determinable by specifically designed living species. Instead, "everything" includes extremely fast-altering integrated or in-depth energy at the elementary levels of energy elements in any unit-value object, item, or mass-energy frame.)}*

→During the progesterone or foundation of the baby Universe in the IEfu, hunting for the 'character attribution' as Absoluteness-status like Pure-energies lay bare /reveal among the affected units of energies.

So, during progression and diversifying functionalities, all units, whether partly or entirely affected or afflicting energies, exhibit trends towards achieving equilibrium or conjoining with 'Pure energies' to regain the earlier stability as the actual Pure status of the Supreme - Universe.

However, their (means, *disturbed units of energy)* infertile attempts cast them aside, either nearer to the periphery or dispersed throughout the IEfu. This futile pursuit of equilibrium status arises from self-generated impediments that stem from 'irregular sliding' activities during adaptation in the spatial pre-embryonic area or the 'IEsu' coverage.

"Nevertheless, this equilibrating tendency acts on time values constancy in a state of the unit/s of energy, capable of ministering steadfastly under prevailing circumstances in the progressing Universe.

This equilibrating practice, whether in the 'IEfu' or remotely at an advanced stage in the '2nd-energy-belt,' can be assimilated with the expression as balancing propensity. It reflects the 'tendency of retention' as steadfastness under the prevailing atmospheric conditions for small or large units of energies as mass-form and energy products, which become significant to stay as units in the mass-form throughout the ruffling Universe."

(By the way, it may hint a little that advanced stages of mass-energy frames, which are referrable to the diversifying status of objects, exhibit exclusively in the matured 2nd-energy-belt zone since this zone occupied ***excrete*** *or residual form of energies from the other two energy belts, that is, 1st-energy-belt and the 3rd-energy-belt.*

These ***excreted*** *releases are the premier gift, which possesses types of fertile energy elements that allow the advanced phases' mass-energy frames to attain or develop into distinctive forms of visible and invisible objects, in addition to the diversified physics values, energy values, or elemental values applications.)*

i. "The technicalities regarding the concept of "equilibrium" and the genetic tendency to achieve a balanced state."

The conventional word-sense 'equilibrium' means equipoise /balanced /stable and likewise to some desiring state for a parametrically invariable steady period.

Accordingly, this objectually recognizing expression 'equilibrium' is considered balanced, equipoised, or stable under some parametric reasonable circumstances, which is acceptable to intelligent species in this biological globe.

Based on the natural way of satisfying the expression 'equilibrium.' At the energy levels, it is to note that all energies, irrespective of their mass state or constitutionally materialized objectual form in different universes or multiverse in the matured 2nd-energy-belt, intrinsically or congenitally endeavor to achieve a state of equilibrium.

From the energy standpoint, the practices of objectually defined equilibrium and related functionality are influenced and derived from the ancestral Pure-energies when energies were at their 'Absoluteness-status' in the Supreme - Universe.

These equilibrating properties began to manifest within the 'IEfu' that carried through generations, even as energies transitioned from Pure status to diversifiable individual classiness, forming the foundation of the actual Universe.

(By the way, it is significant to express philosophically that at energy levels, all detectable and obscure forms of 'physics,' which arise due to diversifying functionalities, were once intact when energies were at the 'Absoluteness status' in the Supreme - Universe. Exceptions arise as comprehensible qualities and manifestations in distinctive manners depend on the specific evolution of intelligent species on this and other habitable globes /natures.)

Hence, based on the biologically evolved intelligent beings, the conditionally articulated word 'equilibrium' is truly an intrinsic property or **'intrinsic-energy-value**.' This

'intrinsic-energy-value' is an integral part of an object, a mass-energy frame, or unit-value energy, revealed during the functioning of elementary levels of energy elements, regardless of any earthly dignified objectual-element or matter-element, constitutionally set in various states of matter like solid, liquid, gas, plasma, etc.

So, this congenital trend to achieve equilibrium or a balanced status may align with any natural occurrence, obeying the energy levels bearing genomic codes. Even though in some cases of equilibrating practices or achieving a balanced status, there are virtually or logically detectable **discrepancies** following the limiting or specific ways of discerning 'dimensional judgments.'

*(By the way, in this context, the term '**discrepancies**' at the energy level may refer to the cause and influence of non-standard, atypical, non-natural, or constitutionally distant values or unsupported non-cognitive energy components that innately procure by observers, measurers, intelligent units such as living beings, alien life forms, non-living entities, and artificially created intelligent systems like AI.)*

Comprehensively, the progressing Universe achieved genetically behaved equilibrating practices due to the intrinsic propensity for corrective strategies among the depleted energies throughout its ruffle energies in the IEfu. Consequently, this 'equilibrium or balancing' practice extended even to the progressing 1st-energy-belt.

Furthermore, this tendency to evolve towards a balanced or equilibrium status has also extended to the 2nd-energy-belt, which is conspicuous in the developed stages of objectual elements.

As the actual Universe gradually advanced, and upon the 2nd-energy-belt slowly poising towards maturation, containing various forms of energies at their inextricable status and phase-constrained proprietorial individuality as matter-energy, matter-product, product-energy, and their diverse releases either as a single unit or collectively as a state of mass or mass-energy frame, all exhibited a common tendency to recuperate genetic features, resulting in a state of equilibrium, balance, or stability under conducive atmospheric conditions.

→Although these earthly represented various states of matter, which appear as objects due to multiple *(energy levels)* physics values applications that are distinctively recognized in the form of solid, liquid, gas, and plasma, are practically multifarious transfigurations of matter-elements or may directly be comprehensively justified as lately developed 'matter forming energies' or 'matter-energies,' all are at their phase-constrained stability, equilibrium, or steadfastness units.

Thus, their equilibrating tendency is not a free form but the involution of many other factors that rely on the total energy values of containments as a unit form, irrespective of any status like solid, liquid, gas, or plasma.

ii. "The Implication of the Equilibrating Factor for the Purest Energy of the Universe."

The congenitally inherent equilibrium factor of the energy elements became one of the bases that grant the extreme to the infinitely fine form of unit-value energy, which is the truly Purest Energy. Their mass state with supreme motion is constrained and unitedly at an equilibrium state in the 3rd-energy-belt.

Here, units of Purest energy may be expounded by the term "Absoluteness-Purest form of energy," as motion elements conclude their concatenation /interlocking with other forms of energy elements throughout the Universe and ultimately become confined in the 3rd-energy-belt. This state of finest form energies is considered Purest energy and may symbolize "0abs∞."

8. Some delineation concerning the ephemeral but indispensable works of energy-element and elemental-value correspond with objectually defined physics and physics-value.

Here, note that the types of proprietorially individual 'energy elements and its elemental values' are innately integrated. Therefore, based on the objectual expression 'ephemerally lasting' on the aspects of proprietorial functionalities, cohesively characterizing energy elements and elemental values can be expressed as one or mono-state and **independent.**

So, analogously, applying the objectual sense and the point of expression 'independent property,' ephemerally-lasting propriety or role of energy elements can also be justified as momentarily **independent** or existing for an extremely short period of 'energy-Time/s.'

This independence span (*unimaginably short*) commenced at the settlement as a functionally dignifiable state of the energy element, and it lasted for when it got disturbed and commenced for alteration. Thereon transits from an earlier signifying property or royal role according to the surrounding conditionalities.

*[{("**Some analytical clarification**"*
*Hence, for example, the above points are albeit remote but somewhat reasonable based on the naturally available 'physics values activities' seen in the sublimation process of a substance, such as the vaporization of frozen carbon dioxide {*that is, CO2 *(solid) to CO2 (gaseous)} at ordinary atmospheric pressure and temperature.*

*Justifying on the point of energy, if dry ice or frozen carbon dioxide is considered an energy level 'energy element,' then the vaporized yields {*that is, CO2 *(gaseous)} due to atmospheric physics values' application will be represented as elemental values (*indicated with a drawing SU24*).*

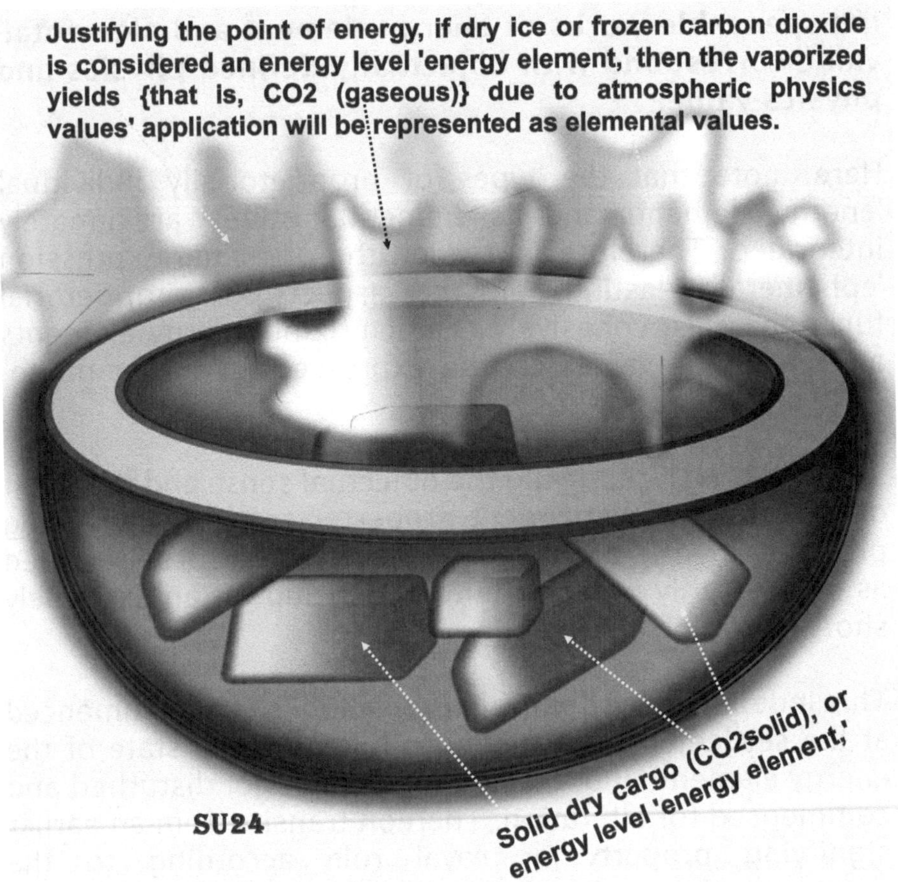

So, it also may mean that assuming the 'solid form of dry ice' unit represents an independent position of the 'energy element.' Then its (energy element) proprietorially independent status will disturbed immediately upon transition from an earlier vantage state.

Also, ideologically speaking, energy levels atmospheric physics to unit-value energy is somewhat similar to the objectual exercise of sublimation to dry ice, which commences due to changes in atmospheric conditionalities like the application of limiting pressure and lowered temperatures.

Besides, it is to say that the evaporation process to the CO2 gaseous will continue for as long as all conditions for sublimation are unchanged to the unit-value solid dry cargo (CO2solid).

However, in the case of the energy elements, objectual roles will seemingly be remote even though objectually deriving physics values yields are the resultant outcome of its integrated works of constituent energy elements. So, 'energy elements' cannot justify applying the self-dependent sense of various states of matter like solid, liquid, gas, plasma, etc. Rather, it can speak that all of their (means, types of state-of-matter) phase-termed distinguishing properties befittingly limited within some reasonable parameters, which is entirely unlike the promptest momentarily changing status of energy elements.

Instead, energy elements are most momentarily functional and consistently keep on changing properties (especially in the matured 2nd-energy-belt*) when undertakIng responsibilities to handle the life of a mass-energy, mass-energy frames, and objects like any of the mentioned 'states of matter' under the conduciveness of atmospheric conditionalities.)}]*

i. A little more clarity on the above from the functional viewpoint of energy during progesterone for the prospective Universe (*that is, in the IEsu*) of the Supreme - Universe, there the expressible diversities were rare and unaccountable mainly for the monopolized role of some 'lowest levels' of **disrupted** 'motion elements.' Or this lowest level of **disruption** may assimilate as energy levels' somewhat most minimal 'asymmetrical activities.'

However, on the applicable aspects, accountable diversities increased with the advancement of progesterone activities, consequentially enriching the discreteness, presumable as mass form units from the earlier status of energies of IEfu.

Because of increasing accountability of discrete 'masses' or as **myriad** units of energies functional roles, gradual enhancement of proprietorial quantability of their constituent activities or integrally serving working components or energy elements, comparing the Absoluteness-status of the assuming units of Pure-energies.

*(Here, the expression '**myriad**' is presumed to be based on distinct terms that are standardly applied to analyze objects using comparable strategies.)*

It means, based on the Absoluteness-statuses of any unit-value Pure-energy, where all constitutionally or integrally set working components or energy elements are at their ineffectually potent position,' or functionally bargaining '**potential abs0^∞**,' so there is no at all possibility of emerging elemental values or energy levels physics values, even though on the point of motion they are neutrally on the sliding move.

*{(The expression '**potential abs0^∞**' here indicates the ability to develop into actuality in terms of function (**potential**), from the standard or objectual abstraction about an absolute zero value (**abs0**) to the possibility of an infinite extension (^∞).}*

Therefore, there's no prospect of the appearance of any countable discrete form of the mass-energy unit, so there is no question about the vulnerability of its integrally unified energy elements for action using its elemental values or energy levels physics values in the 'Supreme - Universe.'

And accordingly, the Supreme - Universe or state-of-Absoluteness can be justified as the 'True eternal constant' explicable for all energy levels measurement purposes, whether intelligible or unintelligible from extreme lowest to highest levels of diversifying activities immediately from the initiation, which even continued after gradually turned to a fully-grown Universe.

→So, remotely, at energy levels' the actions of energy elements and elemental values can somehow be compared in many aspects with the bio-natural and scientifically described objectually presented physics and physics value activities.

For the same, relatively, it is possible to assimilate with the earlier indicated drawing SU24, where the dry ice or CO2(*solid*), if assumed as an evidential physics *(or in a distant manner, the same considered as an acting energy element*) and justified as the source or potentially occupying some form of physics values like the CO2(gaseous) (o*r the same gaseous CO2 considered as an acting elemental value*).

Then, when this physics (that is, *CO2solid*) exhibits in the form of a gaseous state releases (*that is, CO2gaseous*) immediately upon appropriate application of the external atmosphere, the solid form of mass-energy-frame dignified as CO2, which represents a component like physics or the prevailing energy elements, will transform to proprietorially differing gaseous CO2 means, the stand-in /reserved physics values or overall yield that somewhat analogous to the in the depth of energy levels' 'elemental values' of energy elements.

For ease of comprehension, a drawing 'SU25' is portrayed to present the sensible /realizable state of remote links between SU24 and SU26.

Therefore, it is reasonable to say that objectually justified 'physics' (*that is, a unit-value solid form of a mass-energy frame or CO2*) can manifest as the amass of energy elements, whereas 'physics value' (*that is, the emancipation in the form of CO2gaseous from CO2 solid)* will serve as elemental values *(that is, the overall product or release from integrally sets energy elements of the solid form of a mass-energy frame or CO2solid*) that reveal or expose instantly upon disrupting the surrounding conditionality to any *(from most-micro to massive*) discrete unit-value mass-energy.

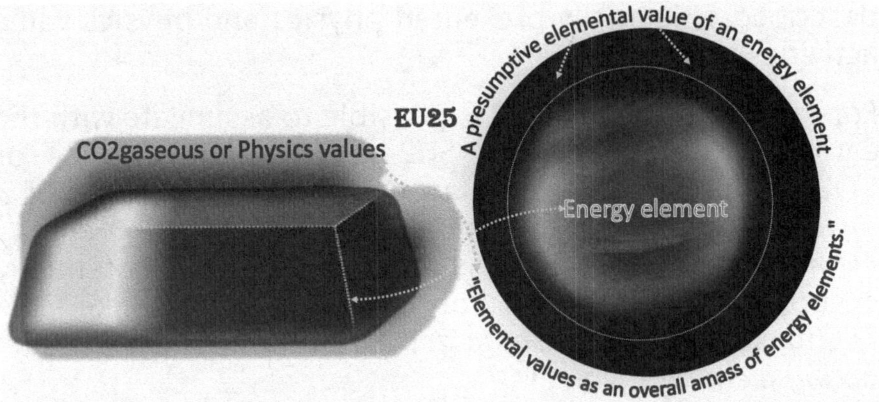

It is reasonable to say that objectually justified 'physics' (that is, a unit-value solid form of a mass-energy frame or CO2) can manifest as the amass of energy elements, whereas 'physics value' (that is, the emancipation in the form of CO2gaseous from CO2 solid) will serve as elemental values (that is, the overall product or release from integrally sets energy elements of the solid form of a mass-energy frame or CO2solid) that reveal or expose instantly upon disrupting the surrounding conditionality to any (from most-micro to massive) discrete unit-value mass-energy.

Indeed, all the above will be much more comfortable to realize regarding the deep down of energy levels' activities if the limits of the ideologically assessing parametric magnitudes are enhanced. Or the same may speak if the judgment quality of the intellect species could extend beyond the contemporary classical ways of understanding following the limits of objectually defined assessable activities.

→Furthermore, the graphic SU26 presented to ease comprehension of the above points, which briefly may be described formulaically as: -

If,

► '**e** = energy-element,'
and its 'exponent-value (**v^9**) represents somewhat like the potentiality of 'elemental values' following the quality and quantity of some state of 'energy element.'

Then (**ev^9**) will convey like: -
The energy element (e) restrained with the strength equal to an exponent-value (v^9) or maybe written like,

► "**ev^9**" = "energy-element and elemental-value" is at their unified mono-state.

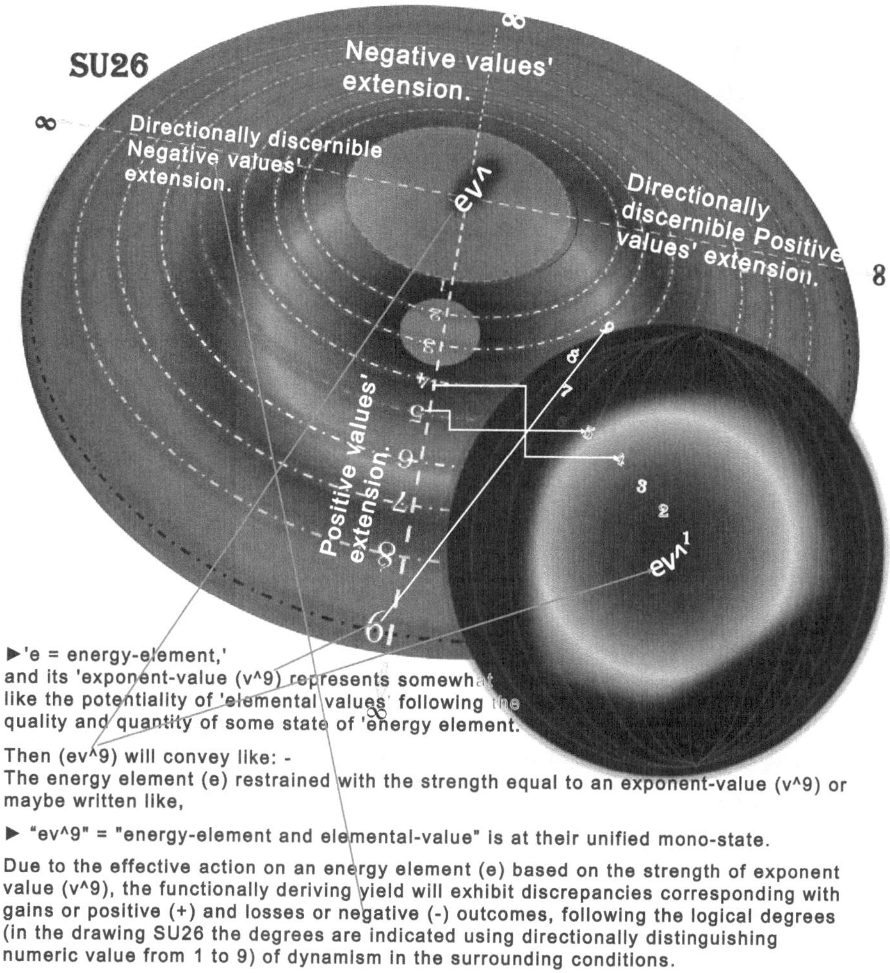

Due to the effective action on an energy element (e) based on the strength of exponent value (**v^9**), the functionally deriving yield will exhibit discrepancies corresponding with gains or positive (+) and losses or negative (-) outcomes, following the logical **degrees** (*in the drawing SU26 the* ***degrees*** *are indicated using directionally distinguishing numeric value from 1 to 9)* of dynamism in the surrounding conditions.

→Now technically, following the objectual sense or analogous strategy between a mass-energy frame that characterizes an object and remotely in the depth of energy levels components, if the exponent-value (**v^9**) represents a specific dignity /identity for an energy element (e) under any particular type /class /category of energies, *(it also may mean that the 'quality' and 'quantity' in terms of the functional property of clench energy of this dignifying energy element (e), must possess potency as elemental values (v) in terms of conceivable numerical value 9, indicated in the drawings SU26 and SU27).*

Then, any variation from the particularity or distinctive level of dynamism, the value of the exponent in the term as nine (9) will alter, which can be distinguished or identified following the consonant of requisites of various intellect species, as thus they seasoned or locked by parametric value due to evolution in distinctive habitable natures.

It means that the exactitude of dignity or acknowledgeable identity based on the **(ev^9)** will keep changing with surrounding implications. It is indicated relatively in the drawings SU26 and SU27 by using numeric terms from 1 to 9, which can again extend to an infinite (∞) value.

Also, if the exponent of elemental values **(v^9)** reaches their final stage, it may mean that the energy element will be entirely depleted and transformed into an (**ev0)** state.

This '**ev0**' state can be considered the lightest form of an energy element, akin to the motion element with properties of speed and neutrality.

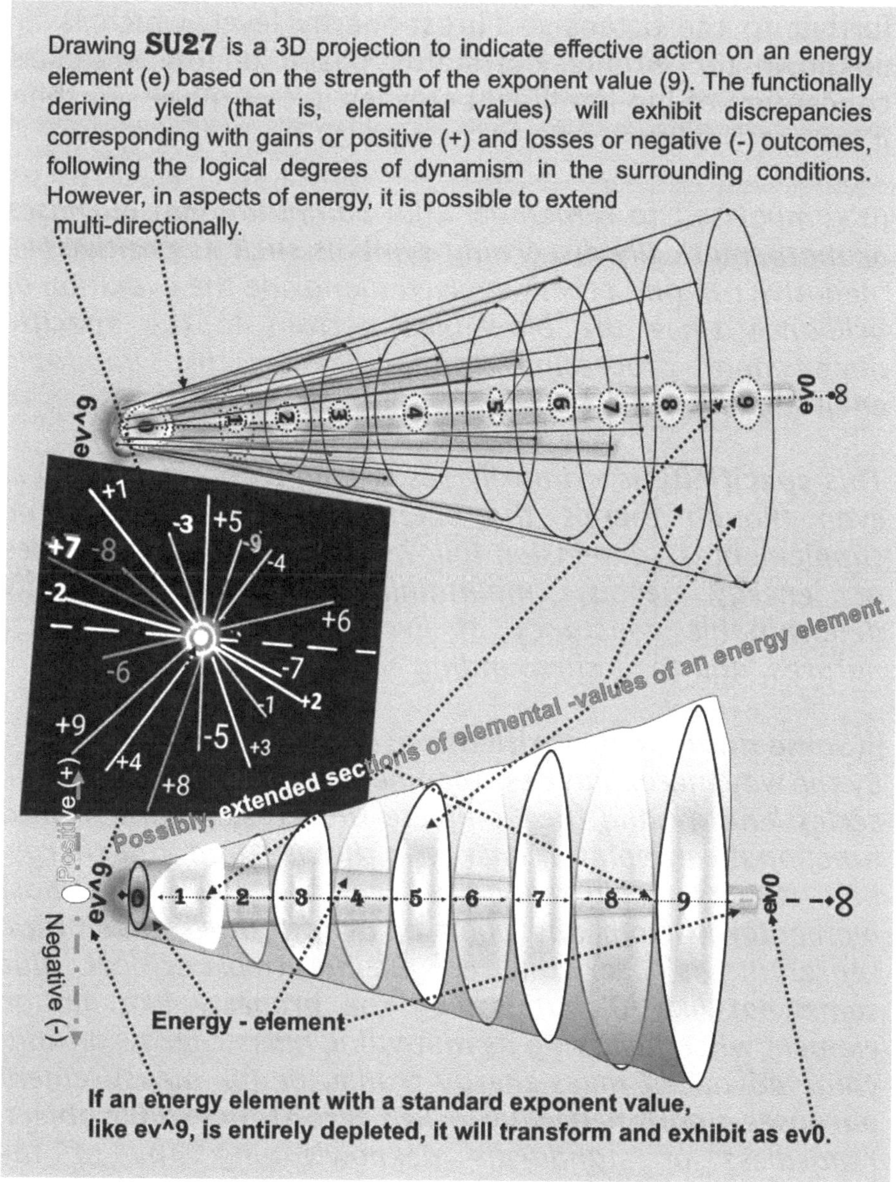

Drawing **SU27** is a 3D projection to indicate effective action on an energy element (e) based on the strength of the exponent value (9). The functionally deriving yield (that is, elemental values) will exhibit discrepancies corresponding with gains or positive (+) and losses or negative (-) outcomes, following the logical degrees of dynamism in the surrounding conditions. However, in aspects of energy, it is possible to extend multi-directionally.

If an energy element with a standard exponent value, like ev^9, is entirely depleted, it will transform and exhibit as ev0.

Objectually or for a distinctive mass-energy frame, this **'ev0'** may be justified as a **decay** stage or impotence within a specific definitive action.

Otherwise, at energy levels' this **'ev0'** can be extended further to the ostensive Purest-energy level, which is an unknown (*or infinite /∞*) status based on the standard conception of bio-intelligent species in this globe. Further, it indicated a little in the drawing 'SU27.'

*(It's important to remember that computational purposes or mathematically discerning symbols such as positive (+), 'negative (-),' and others evolved alongside the evolution of brilliance since the bio-beings pertain to the **specific** dimensional processing capabilities in this habitable environment.*

*This **specificity** is primarily restrained within a 3D space, even though there's the potentiality to conjoin some supplementary dimension following the reality of changes at energy levels, maintaining the parametrically determinable constancy of lives as living-being units, natures, and their surrounding easily assessable items.)*

[{(***Some analytical clarification:***
*By the way, here, maybe remembered from the earlier book series 'Energy and the Complete Universe,' where from a functional viewpoint of energy in the matured 2nd-energy-belt, it tried to illustrate that everything performs most momentarily from any point of **origin** either that origin represents as observably recognizing utmost critical value somewhat like '0' but having the property like motion element when justifying as unit-value mass-energy, or unit-value advanced mass-energy frame, or for measurement purposes a parametrically locked zero-status-stable object, regardless of standardly acknowledging any of the distinctive states of matter.*

*Further, even that the '**origin**' is possible to consider at a most distant like '**ostensive Purest-energy,**' the end or utter quality of any form of unit-value mass-energy or mass-energy frame to any known and unknown class within the matured 2nd-energy-belt.*

Then, the parametrically determining the state of an object (whatever is recognizable like solid, liquid, gas, plasma, etc.) *at most will gradually keep on transforming to different levels /degrees of energy and continue filtering until achieving the criteria as **ostensive Purest - energy**, or somewhat similar to its* (that is, **ostensive Purest - energy**) *property as 'motion element' that neutrally and quickly regulates a mass-energy frame or object.)*

*(By the bye, one thing to remember is that the **ostensive Purest-energies** are the most fundamental regulative essences or the soul of the matured 2nd-energy-belt.*

However, the 'motion element' mainly works for other forms of 'mass-energy units,' mass-energy frames, or objects to their most micro to a massive structural role regardless of distinctive ways of featuring living and non-living categories.

*The motion element is thinner but is a collective outcome of unit-value mass-energy or referable mass-energy frame. Yet, from the viewpoint of significant characteristics such as neutrality and dynamism, the motion element in an object or mass-energy frame somewhat resembles **ostensive Purest-energies.***

*But those motion elements are unlike the far-off finest form of **ostensive Purest-energies** and its universal regulating capacity in this matured 2nd-energy-belt.)}*

*{(Further, one thing to quote is that the appearances of many objectually realizing linguistic expressions like the dead, **decay,** and so on, planted during the evolutionary*

process to bio-living beings, mainly relied on the judgments derived by the individual or their group as a species following the delimitation of the attaining sensing faculties, which determine some level of security for state-of-stableness of the living units as the individual appearance in the place of this bio-habitable atmosphere.

However, in philosophical terms, 'energy' is eternal and juvenile. It (energy) *only transforms from one state to another, fostering an observable belief in living beings as 'illusiveness' concerning the times and events of objects to onlookers in this matured 2nd-energy-belt.)}]*

→Technically, it is ascribable with the equation **'E = mc^2,'** where it is realizable that the entire amount of energy 'E' remains intact, whatsoever its (*that is 'E'*) constituent components like 'm' and 'c' vary from a state of acknowledging value or stance in any form of 'E' that represents as smallish or massive unit-value mass-energy, mass-energy frame, object, etc., throughout the 2nd-energy-belt.

Alternately, resembling the equation **E = mc^2**, if the alphabet 'E' is symbolized by (Eev), which is the overall functional property of energy element (e) and its value that is 'v' (*means, elemental value*) based on the quality and quantity indicated by the possessing coefficient like (^9) then it may be written like:

- **E** (*overall functional property of energy element*) or **(Eev) = ev^9.**

The exponent value (^9), that is, (v^9) of the energy-element (e), has chances of extending to infinite (∞) with the combination of integers (*or something else that can apply to solve the purposes like the integers, which perform various arithmetic operations*) as the requisiteness of various intellects evolved in different habitable natures.

On functional aspects, values (v) of energy elements (e) at different energy levels possibly can extend *(from 1 to 9 or more)*, correspond to matter elements, material elements, 'materialistic elements,' and so on, which are perceptible to bio-intelligent beings.

Moreover, from the standpoint of energy in the matured 2nd-energy-belt, all these matter, material, and 'materialistic elements' can be typically understood as multifactorial formed objectual elements. These objectual elements are constitutionally composed of advanced-stage hybrid /composite energies, distinct from the primordial energies in the immature or infantile stages of the 2nd-energy-belt.

*[{(**Some analytical clarification**:*
One key point to recall from earlier books is that, at energy levels, the objectually satisfying expression of (0) or zero is inappropriate to apply. It is because energy is omnipresent, whether it functions at its unimaginably lowest level of potential or its highest degree of kinetic activity. Thus, the concept of 'ev0' still represents some form of energy, with either a reasonable or unreasonable state of dynamical property.

Also, from the standpoint of energy, one significant point to remember is that the objectual and standard mathematically expressing '0 /zero' does not mean entirely nil /vacant /empty /nothing like senses. Instead, it (0/zero) can justify similar as an absolute-neutral value, or somewhat an unchanging factor or constant on the aspects of any analyzable action and reaction or function, mainly applied to understand perceptible and assessable items.)}

{(Philosophically, a little more may be described for ease the conception of the comprehensive expression of Energy (E) or the alphabet (E), as mentioned in equation form 'E=mc^2' and its infinite standing in the matured 2nd-energy-belt, in terms of proprietorial diverseness, which

conveniently may be applied to determine distinctive expressions like grade, type, class, quality, quantity, and their numerous divisions. These are based on the choices of various energy products, irrespective of whether they manifest as living or non-living items, to specific ways evolved intelligent species.

Additionally, since some amount of energy represents a mass-energy frame or a fully-formed product, such as a living or non-living intelligent unit, species, or other entities, it naturally attains an ***individual*** *status. This* ***individual****ity can be comprehended as a 'zero-status-stable' stance, determined by their constitutional composition, like the bio-grade energy for biologically evolved living species (*including intellectual beings*) in this habitable nature.*

Accordingly, these mass-energy frames or products, characterized by their zero-status stability or cessation, gradually enhance their internal physics together with energy level ***consciousness*** *in coordination with the surrounding facilitative atmospheres, referred to as 'Twig + energy-Atmosphere.'*

*From the viewpoint of energy, the expression '****consciousness****' is illustrated in the book series "Energy and The Complete Universe.")}]*

ii. Now, analogously we may say that **physics** and physics-value, which is limited to the science of this bio-nature, mainly deal with the causes that modify the general properties of objects and their physical status due to specific effects.

→The term "**physics**" is essentially based on **objects** that are associated with bio-intelligent beings.

However, from energy points, **objects**, and their diversifying products exist in many forms, such as 'matter-products shape from matter-energy,' 'material products from material energy,' 'materialistic-product from materialistic energies,' and so on.

To living beings, all those known and unknown objectual products may be conceptually complex and distant from the reliability or **trustworth**iness on the viewpoint of functional properties of primordial units of energies and their integrated energy elements, which have become constitutional parts of the matured 2nd-energy-belt.

*(Here, the expression '**trustworthy**' may imply a sense like 'steadfast' from the perspective of the most transient nature of integrated energy elements or working essences for energy levels activities, especially when compared to a parametrically restraining object or a mass-energy frame.*

Technically, it may be delineated a little more to comprehend that the parameter or parametrically restricting values determine the physical status of an object or the representing mass-energy frame. However, the integrally set energy elements consistently perform to reserve the standardness of parametrically determining an object or representing mass-energy frame under the conditionality of the atmosphere, wherever the object or the representing mass-energy frame resides or traverses through.)

→Whatsoever, correspondingly, 'physics (p) and physics values (v^9)' conveniently can be justified as:

►pv^9 = physics and physics-value.

So, 'physics and physics value' may be assimilated to the earlier defined energy-element and elemental value (**ev^9)** that is: -

{(ev^9) =
energy-element (e) and exponent-value (v^9)}

Or the same can be expressed as the general property of an object or physical substance, where 'physics' refers to the physical status that keeps changing with the evaluation and devaluation of its physics value in the respective providential atmosphere.

Here, 'physics or physical status keeps changing with the evaluation and devaluation of physics value' means if 'p' represents the physics, then the exponent-value '9' can be sectioned in various ways, from 1 to 9 directly, and by using decimals in both positive (*evaluation*) or negative (*devaluation*) options from any standard reference point.

This concept is somewhat understandable in a comparable strategy as indicated in the drawings SU19 and SU24.

Whatsoever the physical status and activities of an object, in the depth of energy at elementary levels of energy elements and the rapidly changing elemental values that constitute a parametrically computable mass-energy frame (*or object*) and its realizable proprietorial roles as a state-of-stableness, will keep on altering with the variation of atmosphere, that is, Twig-atmosphere + Energy-atmosphere.

Thus, this suggests that observable physics or the realizable physical status of an object is primarily the cumulative outcome of the persistent roles played by the constitutional or integrated energy elements within that mass-energy frame or an object, justifiable as a 'state-of-stableness' for assessment through the systems of intellectual beings.

From the standpoint of energy and the underlying practicality of consistent loss and gain under the conditionality of the surrounding atmosphere or atmospheric physics, the physical status of the object will continue to change within a parametrically **poised state**.

This process (*means loss and gain practices*) persists even beyond the limits of that **poised state**, all while adhering to the rules of energy conservation.

iii. "Let's discuss some aspects of the expression 'wave."

Based on the diverse classiness and high standard of properties of energies, the objectually presented physics and physics values could be analogous with the earthly delineated microparticles and their associated energy waves, even though objectually defined quantum and larger particles are composed of a mixture and various configurations of energies.

Or may speak, like the prevailing amount of energies that natural ways evolve to structure a matter, material, and other forms of objectual particles, also built-in with some classes of energy elements, whatever the 'energy released' from this earthly delineated particle, article, object, and likewise in the form of waves, by their inbuilt or foundational components at energy levels.

Or, these energy levels' foundational waves are recognizable by the expression as 'elemental waves' of energy elements cause of elemental values' consistent roles of integral components in an object or the representing unit-value mass-energy frame.

It also may mean, from the standpoint of energy, the yield as **physics-values** of **physics** may be considered the overall outcome of foundational energy elements that emanate in the form of energy levels waves, which is entirely unlike the **standard ways** of recognizing distinct forms of expression for objects or states of matter like solid, liquid, gas, and plasma, like the **wave** is "disruption in a medium that carries energy without a net movement of matter particles."

Instead, from the standpoint of energy, the expression of a **wave** may be defined as "a deviation or disturbance from a benchmark of levels of energy," either that benchmark justified as a reasonably or unimaginable tiniest bit or a large amount of discrete mass-energy, early phases mass-energy frame, or advanced stages referable mass-energy frame exhibits in the form of an object or objectual product.

*{The concepts of "**Physics" and "Physics values**," as indicated in drawing SU24, can be illustrated using an example of a mass-energy frame, such as a unit of solid CO_2* (dry ice)*. In this context, the object or mass-energy frame acts as the "Physics," while the outcome of its overall compositional components, functioning in unison, represents the "Physics value."*

Thus, the "Physics values" derived from the constituent working components of "Physics" can ideologically be justified as a checker and protector. They ensure the security and stability of that object's or mass-energy frame's parametrically maintained zero-status-stable condition.

*(Also, the expression **standard way** may mean fully trusting the perceptible senses or sensing faculties that evolved naturally to a living unit or their group as a species in a place of habitation for specific requirements.)}*

Analogously, 'it may be clarified' energy is universal. However, the genesis as objects, objectual products, and units of mass form energies regardless of their smallish or enormous standing in the matured 2nd-energy-belt, functioning in balancing order in the waves of vast energy Ocean, whether traditionally, and short-sighted judgment assumes as enigmatic.

→Additionally, it is feasible to suggest that the energies emitted from an object can be considered equivalent to the elemental value of an energy element. This elemental value is the foundational outcome that manifests as the physics value of objectual physics, representing the release or liberation of a specific quantity or unit of energy in the form of energy **waves**. This concept is somewhat analogous to the standard unit of frequency measurement in hertz (Hz), an SI base unit representing the 'number of wave cycles passing through a given point per second.'

►If assumed in terms of energy, then it (*that is, wave/s*) implies a quantifiable amount of energy in Hertz form (*or the expression 'Hertz' can be augmented by energy-Hertz or 'eHz' or some other sensible expression*) passing through a given point over energy time.

→In 'SU28,' the drawings are 3D projections that offer an analogous presentation to comprehend the interrelated association between an object or physical entity and the associated physics values, including its constituent energy, deep-down energy elements, elemental values or elemental physics, and elementary-level waves or elemental waves derived from the consistent roles of generated elemental values.

Each unit, whether considered an object or a mass-energy frame, functionally acts as an 'energy -Clock.' The time at different energy levels, or "elemental times," is momentarily derived due to the consistent roles of the elemental values of energy elements, which are the most fundamental factors of any observable or non-observable item.

The release of energy is due to internal elementary-level activities with the assumption that it emanates from a measurable point *(as indicated in the drawings* SU28) rather than from multi-directions of any unit value item, mass-energy frame, or mass-energy. This concept expounded on the energy-time endeavor resulting from the functionalities of elemental values.

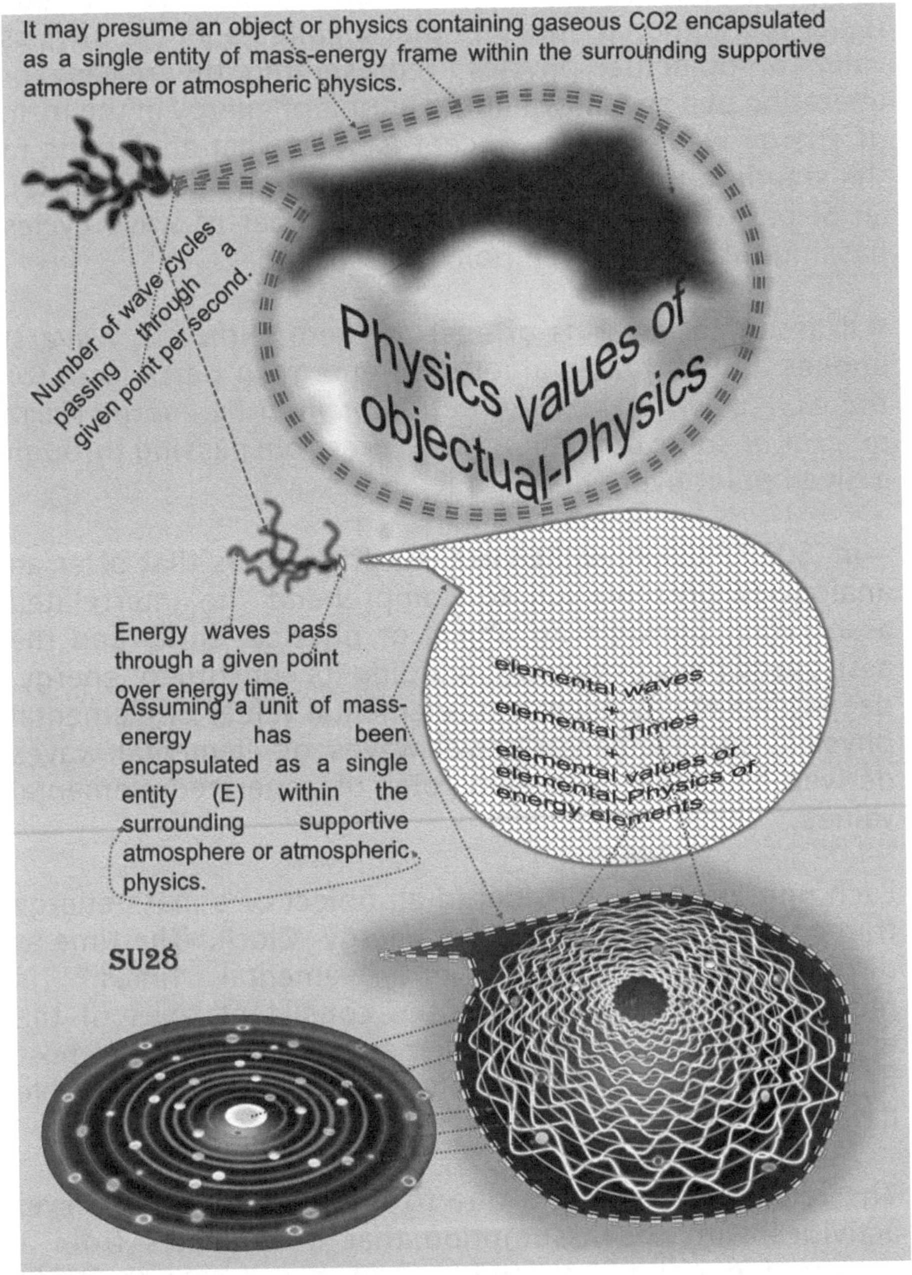

Accordingly, the unit-value mass-energy frame or object is acknowledgeable as an energy clock that momentarily derives all achievable and unrealizable energy times around its peripheral benchmark. It fundamentally differs from the standard time clock utilized by living beings for computational purposes designed to satisfy daily practical activities.

Further, the release of energy at elementary levels can be directly realizable if bio-intellects are willing to push research beyond the standard objectual and mathematical constraints.

These constraints often satisfy only a narrow perspective on the actions and effects of accessible objects, which are fundamentally the overall outcome of units within a mass-energy frame. It encompasses 'not only the analysable components' but also the unanalysable compositional energy levels' working essences or energy elements collectively constitute the entire energy package.

Also, from the standpoint of energy, one thing is to remember from the book series “Energy and The Complete Universe” concerning the aspects of Natural and Universal laws of Physics are co-ordinately performed in this 2nd-energy-belt, regardless of the tiniest to the massive amount of energy as in mass form, mass-energy frames referable or not referable objects of their distinctive classes of energies, all determine the fundamentality of the equation E = mc^, which ensures the totality of energies and their equivalent relation whatsoever their accountable diversifying appearances and roles applying the observable senses of living beings.

→For greater clarity and ease of understanding for a broader audience, the drawing (SU29) displayed indicates the physical areas that possibility of impaction by the physics in waveforms derived by the unit value tidal surge, which is the consolidation of the emanation of ripple and large shapes of waves or speak as a compound composition of multiple types from non-observable tiny to observable large waves. Also, the illustration compares strategies that are both easily accessible and difficult to approach in connection with observable occurrences.

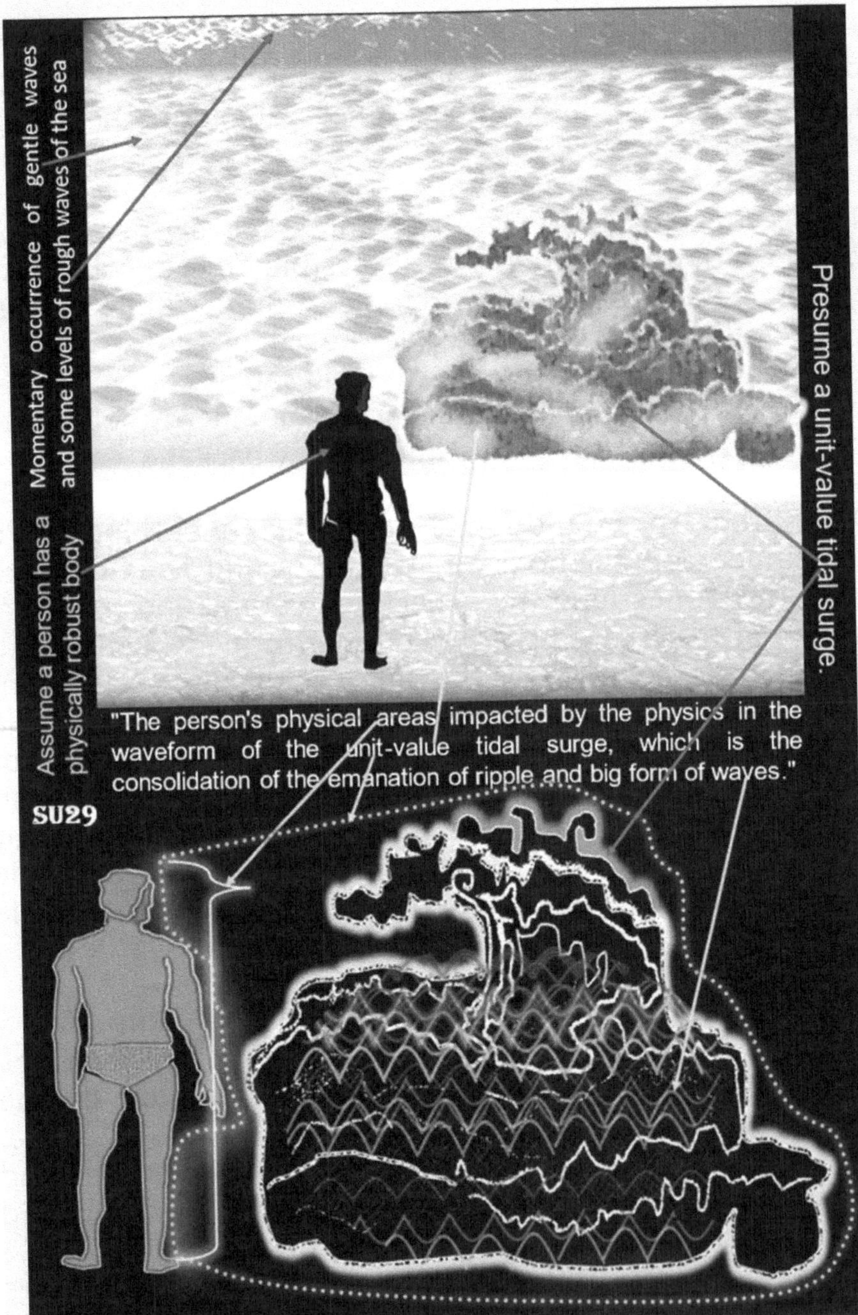

It assumes that most people are familiar with the visible appearance of waves. So, suppose a person has a physically robust body (*meaning someone has a high tolerance for external physical impacts of the momentary occurrence of gentle waves and some levels of rough waves of the sea*) while enjoying the seabeach waves without realizing the potential effects of a tidal surge.

As long as the beach waves are standard and enjoyable, the person's overall physical condition remains stable *(which means the resilience of aggregating organ levels molecular unions of the person)*. However, on a micro level, some of the person's molecules may experience stress, as those molecules are less capable of tolerating the prolonged impact of the reasonable waves corresponding to the other molecules of the overall stout physique.

If that person encounters intolerable tidal waves, the overall internal body physics that secures the perimeter of the physical structure of the physique and the physical stability will be affected over and beyond the tolerable range.

From this example, I think it is reasonable to understand that the tolerable limits emanated due to the internal physics of a child, young adult, strong and then elderly physique or body mass, or the representing mass-energy frames as living beings or their energy levels constituent components, which will defer on a micro level to massive while tackling the circumstances of standard versus tidal-wave's physics values impact of water.

→The earlier drawings SU28, SU29, SU30, and SU31 clarify that the concept waves are explorable on the energy levels. It may be understandable for intellectual beings to recognize how energy waves can operate multi-directionally rather than being confined to the standard three-dimensional perspective that has evolved within the systems of living beings. This concept is demonstrated in relevant sections of the book series, mainly in Energy and The Complete Universe, Part 4.

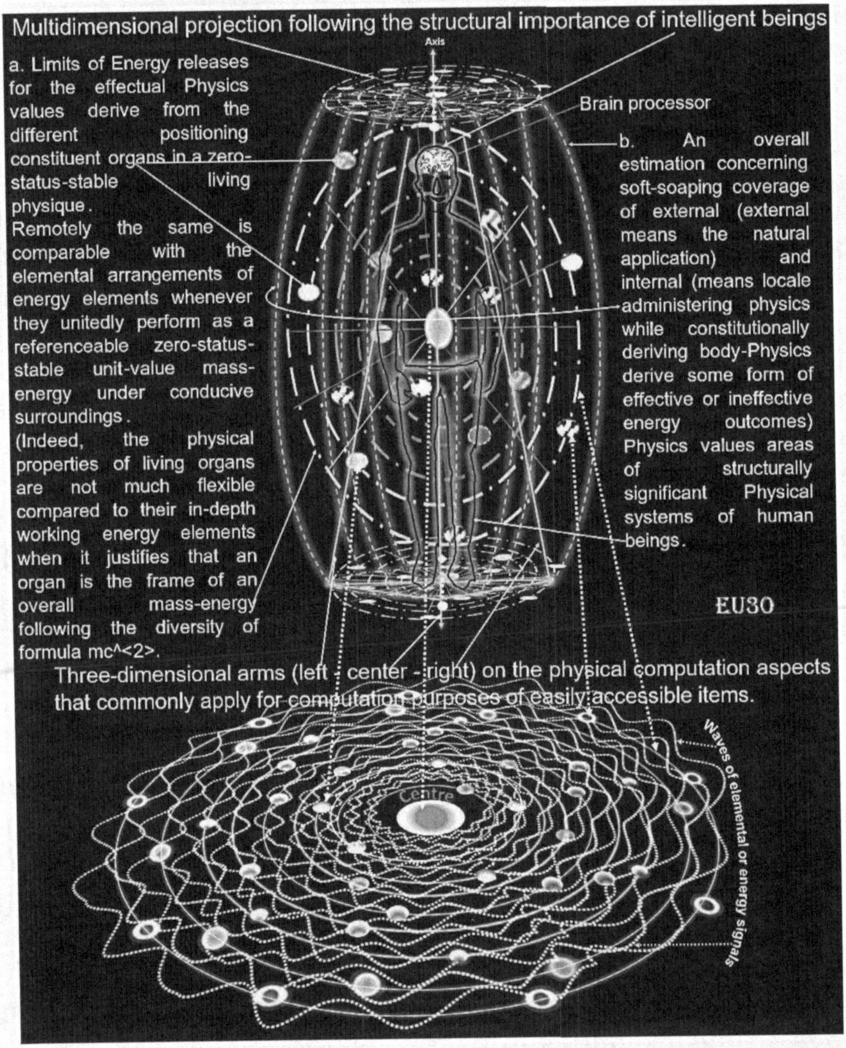

→Furthermore, the concept of waves, as standardly recognized by intelligent living beings, is based on the exposure to particle motion and the direction of energy. These waves are primarily understood as mechanical, electromagnetic, and matter waves, following 3D appraisals. For ease of comprehension, these concepts are portrayed in drawing SU31, along with a few other illustrations featured from the book series "Energy and The Complete Universe," all these drawings provide an augmented explanation of the origin of these waves at the energy levels.

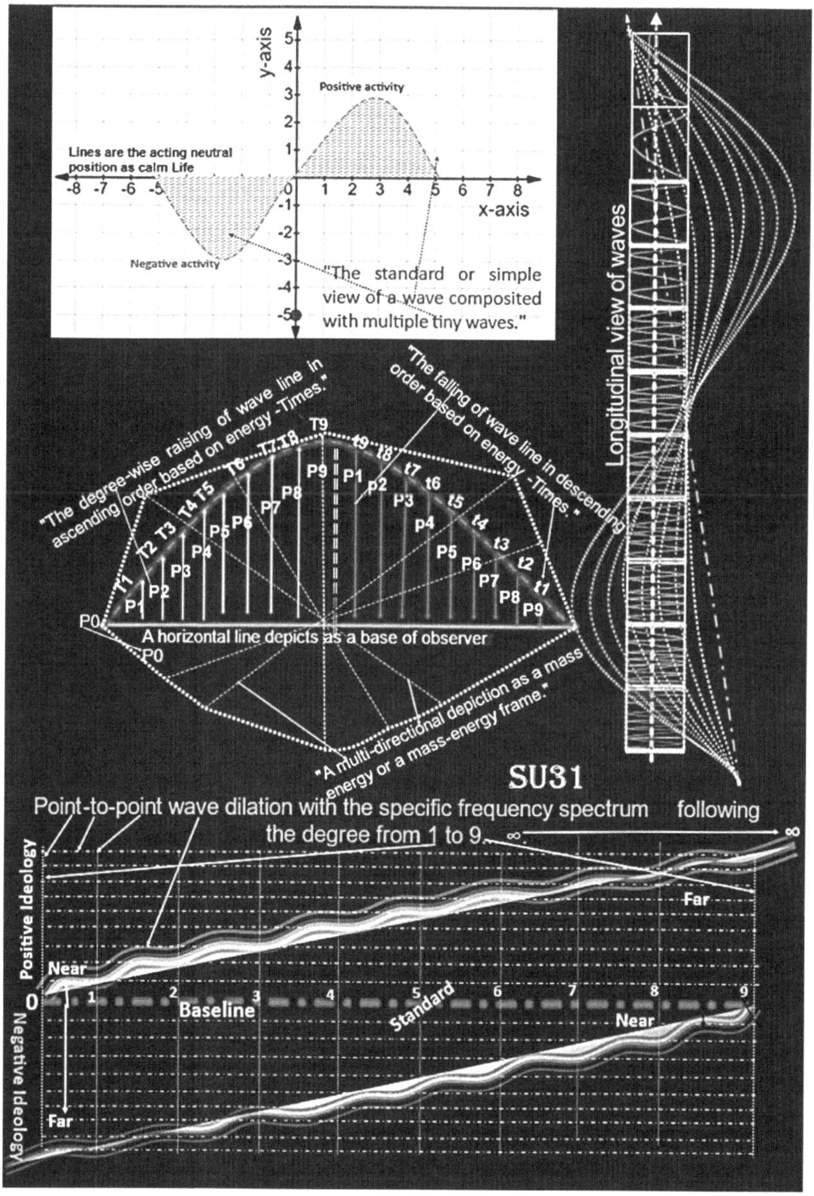

Chapter 11

Let's discuss some significant points of differential value, which is the cause of derivement for gestation /gestatory in the 'IEsu,' as well as a recollection of a few topics on the aspects of such differential value, asymmetry, and the energy-atmosphere in the Universe.

Upon initiating asymmetric activity in a zone of the Supreme - Universe and subsequent breaking of the purity of Pure-energies, it derived to create differential-value among the estrange unit/s of energy due to the disposition of composite elementary component/s of energy.

In general, the term "differential" refers to differences, whether it applies in terms of motion, leverage, or other factors. The expression "asymmetric" indicates a lack of proportionate value.

From the perspective of energy in the evolving Universe, it is justifiable that any asymmetric unit-value energy or its mass form inherently possesses a differential status. It holds real, regardless of whether these differences are considered logically significant or insignificant from the viewpoint of bio-intellectualism.

Further, from the viewpoint of energy and philosophical sense, it may speak of both the differential values and asymmetry, expounded with the state of imbalance or disequilibrium from the originalness as units of Pure-energies or their feature as Absoluteness-status.

Here, it is significant to note that the 'asymmetric' sequence and initiation of the gestatory /gestation process began deep at the energy level within the Supreme - Universe. These asymmetric sequences in the prospective areas as 'IEsu' played a crucial role in leveraging the incubation of the baby Universe within the Supreme Universe and eventually piloted towards becoming everlasting 'differential values' for the matured Universe.

Besides, the creation and consequential evolution of 'elementary levels of physics' or 'elemental values roles of energy elements' gradually became essential to maintain an intrinsic attitude to units of non-pure or disturbed energies for retaining some 'state of balance' or something like the expression symmetry that applied to the Supreme Universe and its Pure energy containment.

By the way, one point to remember from earlier books is that the actual Universe is Truly balanced in coordination with the distinctive three energy belts that are the 1st, 2nd and the 3rd-energy-belt maintaining the eternal constancy or can assimilate as 'true stability' of the 'state-of-Absoluteness' feature of Supreme - Universe *(for this I may be referred to earlier drawing SU9 to ease of one's comprehension).*

i. Based on the reasonability of three-dimensional (3D) assessment practices through the system of a living being, once again, it re-emphasizes that the 'differences in values' among some justifiable unit-value energies at their pure status is the most fundamental factor in the development of gestation /gestatory processes in the 'IEsu.'

This point serves as an outline from a previous book series, where indicated, as a result of the most micro phases differential values, the utmost slow and progressively accumulation and release of free-flowing 'Motional' values begin to occur within the units of Pure-energies in the prospective areas (*that is, 'IEsu'*) for the formation of Universe within the Supreme Universe.

Here, it is noteworthy to remember that the term 'motion' inherently implies a non-ceasing property that contributed to its best effort to create diversity in the aspects of energy or the elementary levels of physics that allowed to lead from initiation, gestation, incubation, and to become a fully developed Universe that is the actual Universe.

→Technically, prolonged consequences with the growing differential- values in addition to 'elemental physics,' that is, elemental values of energy elements, anticipated with expectant /gestatory in the pre-embryonic area (*IEsu*) in the no beginning and no end Supreme - Universe.

Generally, the sense of 'differential value' for objects usually conforms with the difference/s among the comparers. But from the viewpoint of energy in the matured 2nd-energy-belt, the expression 'difference' is the constitutional essence or intrinsic value of any form of mass, mass-energy frame, or object since containing energies in the matured 2nd-energy-belt is a residual form of energies exuded *(mean loss of originalness or advanced stages disturbed pure form energies)* during the progression and structuring the 1st-energy-belt and the 3rd-energy-belt.

At this stage, philosophically speaking, when considering energy, differences may imply that any **loss** of 'elemental-value or energy-value' from unit-value energies afflicts the Pure form of energies or their intrinsic proprietorial feature as Absoluteness-status, resulting in leveraging for some form of **gain.**

Similarly, following a comparable strategy, an object, which is a constitutionally positioned unit-value mass-energy or mass-energy frame, will tend to gain something from other close-by units of energy when facing losses due to some reasons, creating a difference from a particular standing as **symmetry**.

{(***"Some analytical clarification."***
*Here, terms like '**loss** and **gain**' are used as philosophical expressions, often employed to understand some form of deprivation or achievement relative to a standard level of validity.*

Also, one thing to remember following the standard practice of living beings is that, based on the 3D computational practices or judgment of objects, classically, most items can be called back to an equal level or their original state.

But, from the standpoint of energy in the matured 2nd-energy-belt, energy is unimaginably or most momentarily, performing multi-directionally reserving some form of loss or gain beyond the parametric standard analyzing systems of living beings.

So, at energy levels, calling back by balancing the loss and gain to make the expression 'entirely equal' for some objectual items or purposes is elusive. (I think this point is clarified in earlier publications as and when appropriate.*)}*

→The objectually using sense **symmetry** or like balance is somewhat akin /related to the assertion of **'zero-status stability**,' which is in-depth applied for pre- and post-mature objects along with their integral components as energy forms under favorable conditions in surrounding environments.'

{(The concept of **'zero-status stability'** *can be remotely compared to the Absoluteness-status of Pure-energy. In this context,* **'zero-status stability'** *acts as a time-invariant mathematical constant applicable to units of pre- and post-phase mass-energy frames or objects within the consistent and eclectic conditions of the matured 2nd-energy-belt.*

Given the three-dimensional limitations experienced by living beings, for broader computational purposes within the matured 2nd-energy-belt, fundamental conception such as the 'zero-status stability' of pre- and post-phase mass-energy frames, the innate singularity status of Pure energies, and the Absoluteness-state of the Supreme Universe can be effectively utilized.

Therefore, for comprehensive analysis and computational purposes, two fundamental constants like the Absoluteness-status or the innate singularity of any unit-value Pure-energy and an invariant constant for objectual applications can be employed to measure the consistent eclectic roles of various forms of energy and their derived products in the matured 2nd-energy-belt.

Besides, the state-of-absoluteness or Absoluteness-state feature of the Supreme - Universe will retain the eternal fixedness for everything occupied by the actual Universe. (The above point can be decipherable to some levels from the earlier drawings SU1 and SU2.)

Also, it is necessary to remember that for measurement purposes of non-objectual and perceptual items in the frame as a matured 2nd-energy-belt, there is a steadfast /perpetual barrier that can justify as a perfect constant for the 2nd-energy-belt that holds every form of diverse levels of energies and the engendering multi-types products like pre- and -post stages object.

In addition, lots of time-bound and seemingly invariant constancies are available that are predominantly applied by intelligent beings for satisfactory measurement of near and distant observable and non-observable items following the point of measurer/s.

(To simplify the concept explained above with a drawing SU32, besides the earlier depicted SU9 in this book.)}

Further, based on the glory of modern science, I hope it is convincible that the commencement of imbalance is due to disturbing the Absoluteness-status of the lordly unit-value Pure-energies, or the same may speak, the creation of differential-value among the units of Pure-energies consequentially commenced gestation in the prospective '**IEsu**.'

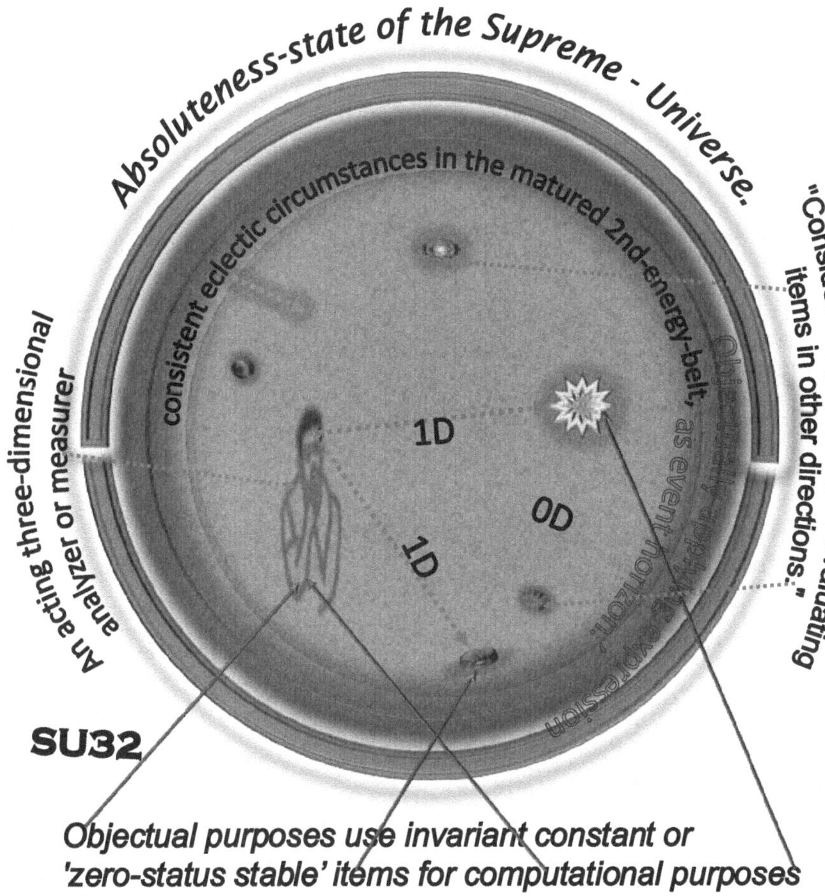

Additionally, the most fundamental level of growing imbalance led to incremental differences in the IEsu, successively driving the progression from the initial phases of gestation through the premature stages of the Universe's development and ultimately achieving the eternal liveliness of its mature form.

ii. Further, for conceptual understanding concerning the differential value, some clarification is provided by depicting the four alphabetical and alphanumerical symbols like **SU, U, E**, and **2bE**.

Based on objectual sense, standardly, all four symbols, regardless of their mono or unison condition, are independent with their differentiable standings concerning the shape, size, magnitude, and also many other physics engendering characteristics, including disproportionate numeric factors like the alphanumeric symbol '**2bE**,' which again may describe as: -

►**SU** = Supreme - Universe = “0absp” = state-of-Absoluteness = Symmetric and bearing ever-constancy in the importance of physics including numeric, magnitude, and likewise = a total or absolute position of **singularity** on the aspects of objectual sense.

(Concerning the expression ***'singularity,'*** *I think have already been clarified in one of the book series "Energy and the Complete Universe")*

►**U** = Universe = (≈0) = individually and comprehensibly symmetric, but from a constitutional and functional viewpoint, 'U' is engrossed with differential values.

►**'E** = Energy,' in this context, the word 'Energy' is considered as the containment throughout the Supreme Universe (SU) as well the actual Universe (U), rather than the usual conception that analyzes based on this habitable nature, where 'Energy' is often discernible in the form of numerous objects and objectual products or matter and **matter-product,** etc., and vice versa *(means convertible manner of their appearance as objects and energies)* within the matured 2nd-energy-belt.

*{("**Some analytical clarification.**"*
To *simplify the usage of the terms mentioned above, when discussing energy products such as* ***matter-product****, material-product, materialistic-products, etc., it's significant to consider that their functional outcomes, whether productive or ineffective, are ultimately characterized as forms of energy or other related entities.*

Additionally, it is important to recall from earlier books that the concept of the "actual form of energy" has been explained in the context of its consistent activity within the matured 2nd-energy-belt. Whether it is discernible through perceptible senses limited by accessibility or dismissed due to its impalpable extreme vibrant celestial roles, where the sense 'energy' is often described using terms like 'nothingness,' the lowest degree of 'potentialism,' or extreme 'kineticism,' all of which lie beyond the limits of our inherent sensing faculties.)

(Analogically, it may be clarified based on the intellectually depicting hypothesis about the expression 'relativity' that is, the entire things /objects in the 2nd-energy-belt truly relative on the aspects of their appearances, as there is no competent absolute or unconditional snare /frame available for reference.

Since from the standpoint of energy, roles of momentary diverseness overpower throughout the matured 2nd-energy-belt, maintained distance the factuality of observable reality that mainly delimited to some discernable roles of matter-product, material-product, materialistic-products, etc., to specifically structured living beings.)}

►**2bE** = Any form of unit-value-energy in the matured 2nd-energy-belt must have some form of potentiality, somewhat can be characterized as the 0(+/-)^∞, where 0(+/-)^∞ mainly applies to objectual senses but can be extendable when analog on the aspects of diverse types of energies.

It means that the symbol '2bE' represents a unison feature as a unit-value like objectual sense as '0', nil, nothing, etc., but possibilities of enlargements on the point of energy with considerable numbers of the wide-ranging amounts of positive (+) and negative (-) notions, matching on 3D assessment and opinion from an individual to their species as living beings, within the entire amount of energies in the vast energy-Ocean of the 2nd-energy-belt.

(To comprehend the 0(+/-)^∞, it may conveniently refer to the earlier drawing SU17 and the new SU33.)

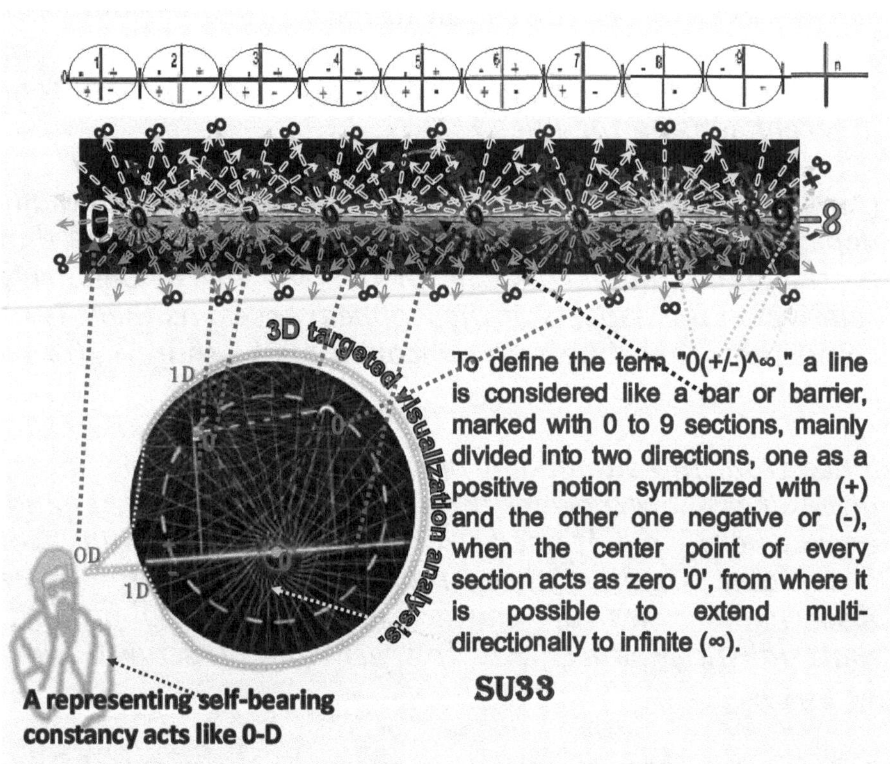

SU33

By the way, following the three-dimensional perspective, the '0(+/-)^∞' analytically can explain that if the symbol '0' represents an assessable **bar, barrier**, from where roles of energies commenced their inaccessible journey towards positive (+) and negative (-) path and possibly can extend to an unknown value like infinite. However, one point to cite is that actual energy performs multi-directionally, entirely unlike that devoted through perceptible standard three-dimensional roles to determinable objects.

*(From the standpoint of energy, one point to remember is that the observably discovering symbol '0' may not approve that the expression like a '**bar** or **barrier**' will act as nil or nothingness senses where invisible energy levels can perform micro momentarily, yet the objectual items and energy conveniently clarified on analogously to ease of comprehension of modern living beings.)*

→ *(Here, to encourage comprehension exceeding the limits of classically developed sense to living beings, one significant point is to remind that visible and invisible objects may not exist everywhere in the 2nd-energy-belt.*

However, the actual form of energies is covered beyond the confinement of visible and invisible objects, having no shape, size, color, or odor, yet it (energies) play crucial roles in multi-type creations and the existence of life, as determined by living beings.)

Whatever, all the assessable units will be typically limited due to the natural way evolved sensing faculties to living beings while discerning and analyzing something through the delimitation of the brain processor of an individual like the massive, macro to micro status as appropriate on parametric marking. Either that (*pursuing through parametric marking*) is recognizable as a satisfactory expression like symmetric, asymmetric, equal, or difference, etc., since the naturally evolved living systems *(besides, for their **purposes** use non-natural creations like AI, i.e., Artificial intelligence)* are deficient in many aspects analogous to the containing in the vast energy-Ocean.

*{Here, the expression "**purposes**" is mainly utilized for specific contexts where intellectual findings connect to their intended goals or desired outcomes. So, the essence of "purposes" refers to how the discoveries or products of intellectual effort are aligned with or directed towards some objectives. The **relative relations** suggest a correlation between what is discovered* (the subjective product) *and what aims to achieve* (the target desired*)}.*

So, they (*living beings*) can induce, assess, and analyze analogously based on built-in or self-bearing constancy as an intelligent living unit or comprehensiveness outcome of their group as a species for the diverse qualities of energies and their product form in the vast or infinite areas of the matured 2nd-energy-belt.

Moreover, the aspects of componential values as diverse types of 'energy elements,' the proprietorial features of units of energy such as (2bE) are far from comparable with any unit-value Pure- energies in the Supreme - Universe, unit-value energy in the IEfu, unit-value energy in the 1st-energy-belt, and the Purest energies in the 3rd-energy-belt.

In essence, it means that on the point of 'elemental characteristics' of energy elements or working components, Pure energies of the Supreme - Universe are more comprehensively sufficient than the units of energies held by other individual energy belts in the matured Universe.

iii. The above complex and abstract concept involving energy and philosophical notions of symmetry, asymmetry, and differential values that are conveniently utilized for the birth of the actual Universe may be simplified and illustrated like: -

►The initiation of asymmetric activity in a zone of the Supreme - Universe leads to the breaking of the purity of Pure-energies, which means disruption of true original

Absoluteness-status, resulting in the creation and incremental upsurging of differential values besides elementary levels of physics in units of energies.

►The expression asymmetry is apprehensible as a lack of proportion or balance. When energy becomes asymmetric, it creates differential values, meaning there are differences in the properties of energies, specifically on the prospect of motion and leverage.

►From the energy and philosophical perspective, asymmetry and differential values represent a state of imbalance or disequilibrium. It also suggests that these imbalances began deep within the energy level of the Supreme - Universe, initiating a gestation process leading to the formation of the baby Universe within it.

►The creation of differential values leads to the gradual accrual and loss of compositional integrity *(i.e., elemental integrity of energy elements)* in units of energy within the prospective areas of the Universe or IEsu. These changes happened during the gestation process in the Supreme - Universe, where the units of Pure-energies *('Pure-energies' termed Absoluteness-status)* begin to differentiate and lose their initial symmetry.

►From the standpoint of energy, everything is relative in the matured 2nd-energy-belt since there is no absolute reference frame due to the consistent practices of **loss** and **gain**, making all entities in the matured 2nd-energy-belt Truly relative when judgment performs following an individual intelligent unit or their group outcome for comprehensive assessment. This concept aligns with the law of conservation of energy, stating that energy cannot create or destroy, only transform.

*(It's important to note that in standard usage, the term "**loss**" is often equated with destruction, while "**gain**" is*

understood as the creation or emergence of something new, derived from a previously validated state.

For different types of energy, 'loss and gain' can be justified by changes from an endorsed perspective or may indicate a reciprocal response.)

Additionally, based on the intellectualism of living beings, from the standpoint of energy, the units of mass-energy frames or objects, typically identified by their parametric order, maintain a **relative relationship** within the stipulation of the time frame of their uniform identity. Therefore, if the standard scaling practices through parameters are narrowed down or ascend, it will act on relative measurements.

Something like a monad is a simple, indivisible unit of existence. In the case of living beings, it is comprehended as an individual organism. For example, a single human represents a monadic unit. But, when considering humans as a group, collectively, it identifies that each individual contributes to the overall human species. Similarly, a monkey represents a member of a different species and has distinct characteristics that differentiate it from humans and other species.

Again, the qualities and characteristics of different species, such as monkeys and humans, highlight the diversity in the biological world. Each living **unit** and its collective group form a species with unique attribution that sets them apart from others in a distinctive way of recognition. For instance, humans and monkeys differ significantly in many aspects, such as cognitive abilities, physical features, and behaviors.

In regards to the above, from the standpoint of energy, fundamentally, the perception and judgment over qualities and characteristics is an engrossed outcome when represented as a small or large living **unit**, which developed interiorly at energy levels to an individual in the evolutionary process that follows the conditionalities of atmosphere *(mean, Twig-atmosphere and Energy-atmosphere)* in the place of habitation.

However, non-living items, namely rocks or water, are primarily different from living organisms. Although non-living entities do not display similar life processes, they are recognized as composed of various 'states of matter' that also originate from fundamental forms of energy.

So, it will be authentic to say that recognizable and unrecognizable, all diverse forms of energy in the matured 2nd-energy-belt conversely perform their roles as energy and objects.

Whatsoever the distinctiveness for recognition implies following the intellectualism of living beings, standing on the parametrically measuring identities as matter, material, materialistic items, livings or non-livings, all are interconnected through the essences of energy, constitutionally deriving by energy physics, abiding by the universal laws of the matured 2nd-energy-belt, natural laws, and fundamental levels of physical laws. "Moreover, the conduciveness *(or this 'conduciveness' may be justified by the word adaptability)* of conditional atmospheres go in with significant roles in a place of appearances, accommodation, or traversing."

►Since there are no ways to observe the Supreme - Universe, the actual Universe, and the True form of energy, thus in essence, my related book series mainly describes analogously, relying on contemporary perceptible items and the logic of metaphysical or theoretical frameworks to illustrate how the Supreme - Universe gives rise to differential values and asymmetry at the Pure form of energy level, which then leads to the formation of the Universe, thereon progressively appearances of the proprietorially individual three energy-belts.

In addition, it uses geometric and symbolic representations to illustrate how these concepts interact and evolve systematically over eons, particularly emphasizing the lack of absoluteness but the inherent relativity of energy and object, matter, material, and likewise items, which recurrently appear and disappear predominantly in this matured 2nd-energy-belt.

iv. [{("Clarification about the word 'unit' that uses many places."

{One point is to remember that the standard term "unit" (primarily used for objectual purposes*) can have various meanings depending on the context. The specific meaning of "unit" is determined by the field or context in which it applies.*

However, when considering 'Pure-energies' and thence advanced stages of energies before the appearance of objects and objectual products, it is challenging to define that an amount of energy can serve as a "unit."

In the case of pure and advanced stages of energy, the term "unit" may not refer to an "amount of energy" in the conventional sense typically used for measurable purposes, such as yield and utilization in practical applications. This is because such stages might involve concepts like the vacuum, emptiness, theoretical constructs, or fundamental constants, which do not always align with standard measurements for energy.

Therefore, in this book series, the expressions 'unit,' 'unit-value,' and similar expressions are applied to describe energy levels' activities, which are hypothetical statuses. Because energies are 'omnipresent and omnipotent' and the roles of energies are the most instantaneous, and perform exceptionally fast, functioning beyond realization through the standard senses that developed in the intelligent species following evolutionary technique and due to the benevolence of this habitable biological environment, which primarily allows to perceive and interact with a limited range of accessible phenomena.}

{For more clarity about the term 'unit' and ease of comprehension, drawings SU34, SU35, and SU36 with their respective frames are portrayed using analogous strategies and descriptions.

→In drawing SU34, some topics and subtopics are marked for ease of comprehension.

a. "A football is made from various materials, allowing it to be pressurized to maintain its spherical shape and withstand internal and external forces during play."

➢A football is known as an objectual unit, an air-filled sphere.

➢The air-filled sphere contained numerous microparticle units in various forms.

(I think everyone knows the specific purpose of the design and inflating of the football, which mainly tackles the reasonable applications of external and internal physics in the conduciveness of the surroundings.)

b. Instead of containing multiple units of microparticles and free-flowing energy, it contemplates that the football (that is, an unison state) *consists of similar numerals of units of individual mass-energy or mass-energy frames that consistently function within the confinement of its solid spherical shape.*

c. A reasonable and comprehensive understanding of waves emerges when we consider the interactions of countless microparticles, units of mass-energy frames, or the essential energy elements and levels that constitute a mass-energy frame, similar to how an air-filled football is composed.)}

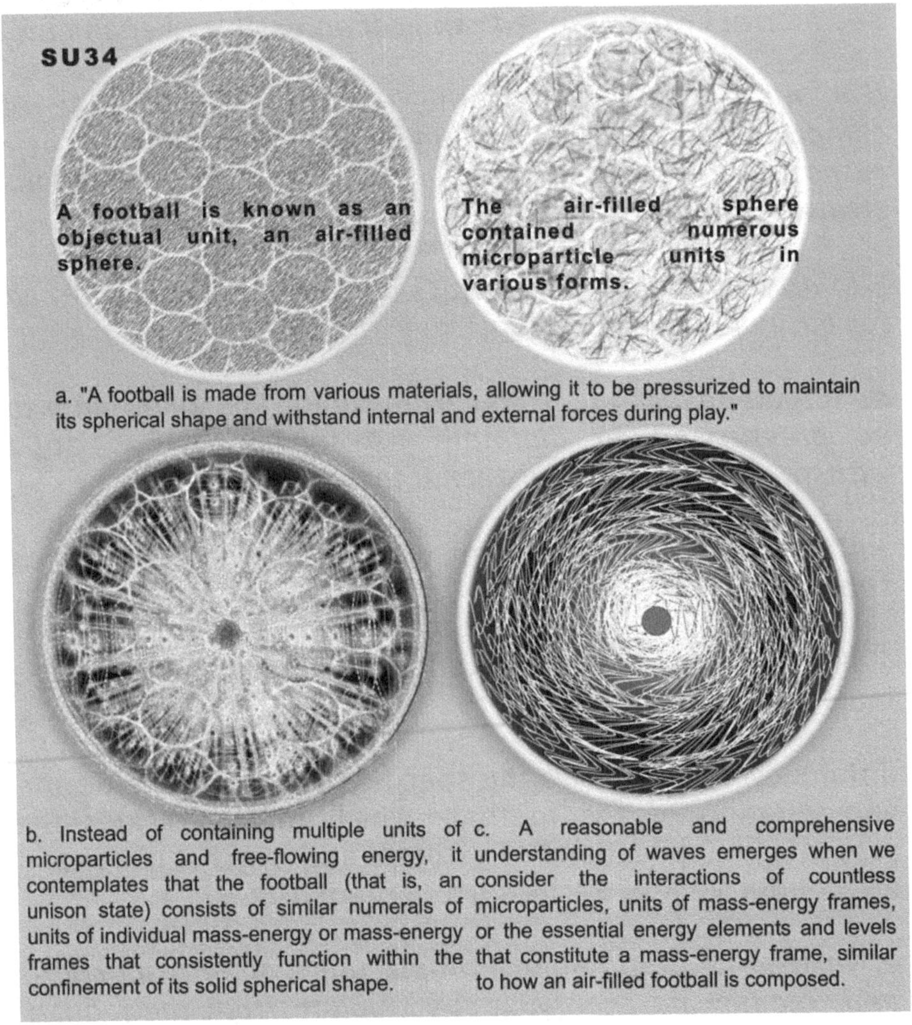

→In drawing SU35 the topics and subtopics are marked for ease of comprehension: -

d. This drawing provides a sectional view of point (b) in SU34, offering a detailed understanding of 'mass-energy' frames as suggested by the equation mc^2. In this context, (m) represents mass, while (c) stands for the motion-deriving element or motion element, constitutionally positioned as a unit-value frame or squared (^2). This visualization helps clarify the relationship between mass and energy within the framework of the equation.

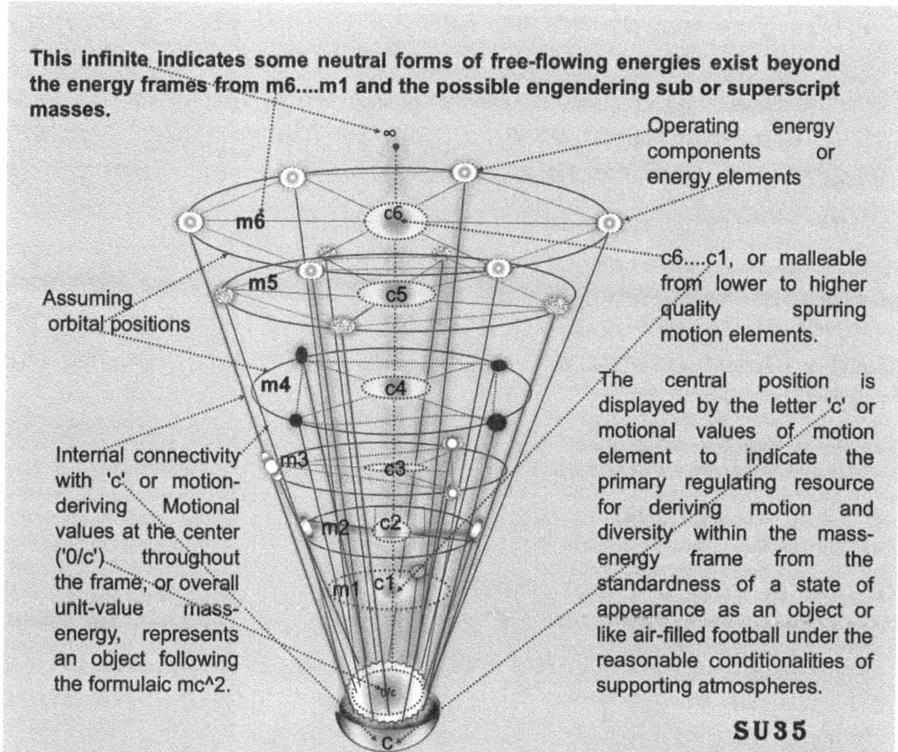

SU35

d. This drawing provides a sectional view of point (b) in SU34, offering a detailed understanding of 'mass-energy' frames as suggested by the equation mc^2. In this context, (m) represents mass, while (c) stands for the motion-deriving element or motion element, constitutionally positioned as a unit-value frame or squared (^2). This visualization helps clarify the relationship between mass and energy within the framework of the equation.

➢*This infinite indicates some neutral forms of free-flowing energies exist beyond the energy frames from m6....m1 and the possible engendering sub or superscript masses.*

➢*Operating energy components or energy elements.*

➢*Assuming orbital positions.*

➢*c6....c1, or malleable from lower to higher quality spurring motion elements.*

➢*The central position is displayed by the letter 'c' or motional values of motion element to indicate the primary regulating resource for deriving motion and diversity within the mass-energy frame from the standardness of a state of appearance as an object or like air-filled football under the reasonable conditionalities of supporting atmospheres.*

➢*Internal connectivity with 'c' or motion-deriving Motional values at the center ('0/c') throughout the frame, or overall unit-value mass-energy, represents an object following the formulaic mc^2.*

→*In drawing SU36 the topics and subtopics are marked for ease of comprehension: -*

e. Based on the standard or 3D (i.e., 0, +ve, and -ve directions) *conception of a full-wave cycle, this wave is considered a unit value occupying a certain amount of energy for performing functions as a micro or large state of its appearance.*

➢*Mark as a distinctive wave that represents a full-wave cycle.*

➢*"The center point where the directional change from the positive and the negative cycle of the wave marks deviate."*

➢*An amount of occupying energy.*

(Based on the standard concept, a realizable single full-cycle wave can be portrayed as a quantity of energy possessed.)

➢*Smooth full-cycle energy waves.*

➢*Unsmooth full-cycle energy waves.*

➢*'Some state of the intricate pattern of waves.'*

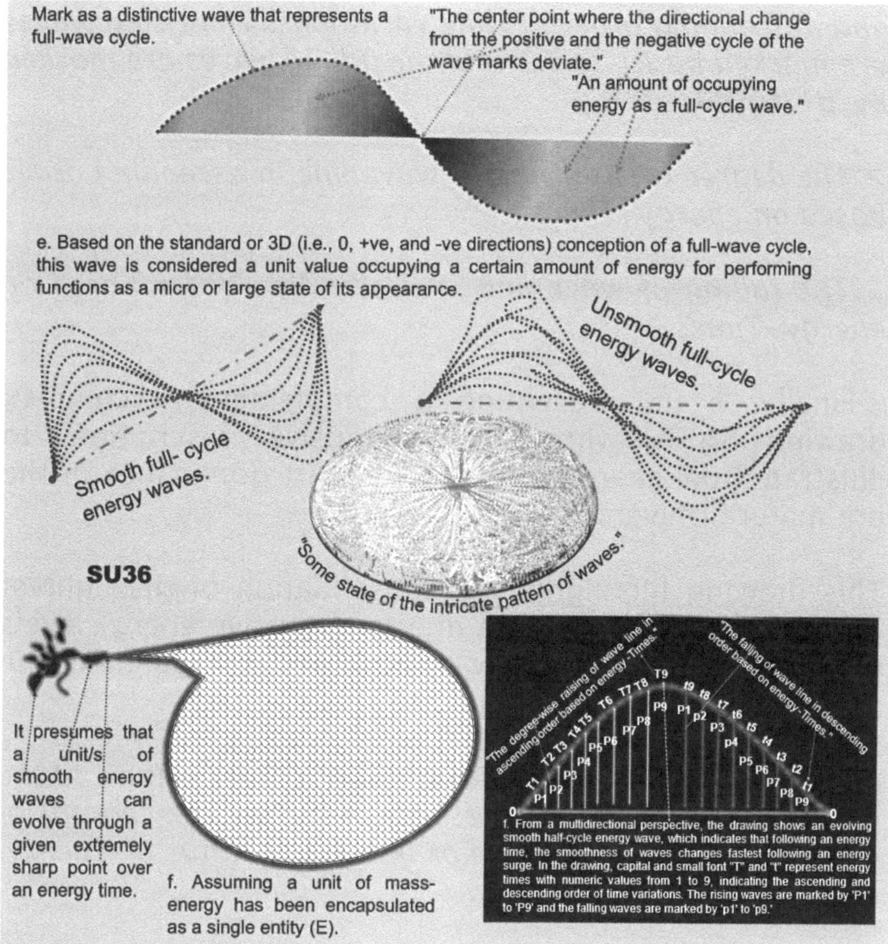

f. Assuming a unit of mass energy has been encapsulated as a single entity (E).

➢*It presumes that a unit/s of smooth energy waves can evolve through a given extremely sharp point over an energy time.*

g. From a multidirectional perspective, the drawing shows an evolving smooth half-cycle energy wave, which indicates that following an energy time, the smoothness of waves changes fastest following an energy surge. In the drawing, capital and small font "T" and "t" represent energy times with numeric values from 1 to 9, indicating the ascending and descending order of time variations. The rising waves are marked by 'P1' to 'P9' and the falling waves are marked by 'p1' to 'p9.'

➢*"The degree-wise raising of wave line in ascending order based on energy -Times."*

➢*"The falling of wave line in descending order based on energy -Times."*

→Finally, I am presenting a complementary complex drawing SU37, without an extensive description, to illustrate how energy level waves exert dominance in the pre-mature Universe.

This drawing highlights the consolidation of the lightest form of energy at the core due to extensive energy levels **physics**, which eventually contributed to planning the prospective 3rd-energy-belt.

(In this context, one significant point to note is that, at the earlier stage of the progressing Universe, the concept of appearance as a unit-value or individual status of 'energy-mass' is highly momentary.

*Therefore, the overall constituent energy elements and their consistently derived elemental values from that fleetingly appearing unit-value 'energy mass' represent the energy **physics** at that level.)*

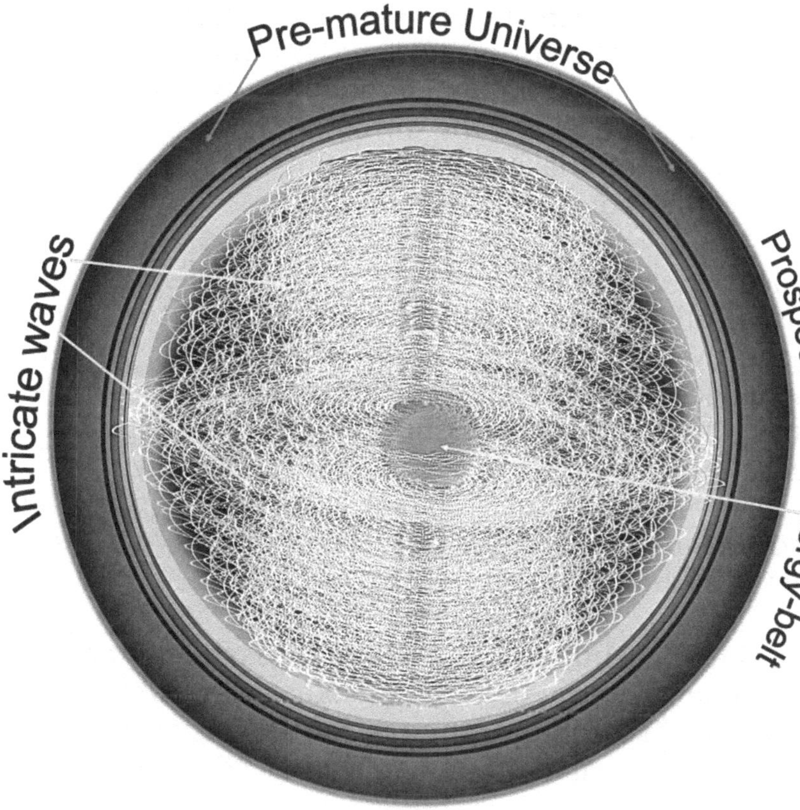

SU37

1. **"Brief illustration concerning the recurrent relation of energies and their constructed energy products, and like the objects and their diversified sub-products in the matured 2nd-energy-belt."**

From the energy standpoint, the expression **relativity** is also appropriate for units of energies at their individual or collective oneness status throughout the **2nd-energy-belt**.

Based on the standard conception, it may simplify thinking of a bottle filled with distilled /filtered or purified water. Whether someone considers a single drop or the total contents of the bottle, the water is still in distilled /purified /filtered form, which means that the characterization as purified water will remain the same, whether in the form of a single drop or the combined volume of multiple drops in the bottle. Indeed, given the different states of matter, the accountability of the bottle is distinct from the water in satisfying the requisite requirements.

→Hence, the authenticity of **relativity** prevails in all forms, statuses, or designs of energy - products in this matured 2nd-energy-belt, irrespective of known and mysterious objectual levels of micro and macro scale particles, objects, matter, their various working releases like signals, and likewise are hypothetically, and even practically reversible or transformable to their distant state of origin as energy or advanced stages '2eB' or units of energies in the matured 2nd-energy-belt.

This point also suggests that the possessed energies in the matured 2nd-energy-belt retain exactness, whatever the perceptible discernment emphasizes expansion or inflation and contraction or deflation following some dimensional discretions.

→Hypothetically, it is rational to suggest that the previous status or individuality of an object and its product might **re-appear** or **re-emerge** in various forms within the vast energy sectors of the cosmos, including planetary systems, either as a replica or in a similar manner. However, it *(visible, invisible object/s and their diversified status of yields)* will not be thoroughly equal in all aspects, considering its individuality or mass unit/s derived from energy and the value or significance of 'energy-time.'

Since in any form of 'mass-energy,' the engendering energy-times, that is, the 'elemental Time' in the depths of unit-value energy or at the elementary levels of 'energy-element's time' has a one-way trend to advance to the final state and eventually converge with the 'ostensive Purest-energies' or something analogous to the mythical concept as the **soul** of the 2nd-energy-belt.

*(Note: The term '**soul**' is defined in the earlier book series "Energy and the Complete Universe.")*

This **reappearance** or **re-emerging** technique performs mainly due to the harmonious re-generation of the 'motional-values' of motion elements deeply set in the unit-value energy, regardless of mass or objectual forms. It *(unit/s form as an 'energy Clock')* involves the change in status or identity of products as a representation of energy flow on 'energy-times' or 'elemental times,' which depends on the deviation from an assumed expression like '0' or zero-energy point or a state of complete balance /stability under **standard** conditions. It can be considered the **standard** parametric significance under specific circumstance/s in a given environment.

→From the standpoint of energy and the vast 'energy - Ocean' as the matured 2nd-energy-belt, the simple way it may speak, every distinctive or indistinctive energy product or object and their progenies endure the feature of relativity to their state of individuality as they are not entirely equal in all aspects.

Rationally, in this nature of bio-intellect science, **relativity** is primarily manifested in the various forms of advanced or conceivable matter and material products based on the **variation** of conceptual parameters. This **variation** originates from a hypothetical 'zero-base-point' to justify a similar approach, whether it involves a complete or partial extension of logical ideology.

So, the objectually presented symbol '0' acts as a 'zero-base-point,' somewhat reasonable to speak as a notionally 'absoluteness' stands on distantly related to the 'Absoluteness-status,' which principally applied for most antecedent /earliest unit-value energies at their Pure forms or units of Pure-energy.

→Furthermore, for the matured actual Universe, based on the expression 'technicality' and the standpoint of energy, the interconnection of energies and their resulting diverse energy products, such as various objects and their derivative products in the developed 2nd-energy-belt, can be comprehended through the unremitting interaction and transformation processes that characterize the dynamism of this cosmic domain.

In the matured 2nd-energy-belt, energies exist in various forms. Every form of unit or unit value 'mass-energy' has an eventuality to give rise to or promote distinct forms of energy products or transform them into another state of energy. The products as mass-energy frames, much like objects in the physical world, are manifestations of the underlying energies that constitute them.

Since energies at their unitary state mass and mass form frames interact with one another under some surrounding conditionalities, they encounter transformations that lead to the creation of diversified energy products relying on the supportiveness of the surrounding conditional atmosphere. These products as unit/s possibly range from simple energy forms to complex structural entities, which possess distinct properties and functionalities following the derivement of internal and external physics or elemental values applications.

(Here, based on the specificity of the intellectuality of living beings in this biological world, it's significant to understand that recognizing, acknowledging, and computing energies in their small amount, unit, or mass-based form as an observable object relies on discerning practices that compare the surrounding accessibility, or inaccessible range possibility with hidden conditions like infinity.

Something like a congenitally visionless and clear-sighted being needs a different way of teaching to know or discerning some states of matter like solid and liquid.)

The connection between different types of energy and their resulting products, whether observable or not, is a recurring theme. These energy products or objects emerge due to multiple functions *(like elementary phases elemental-values roles, elemental-physics or energy levels physics)* from various energy levels to their constituent 'energy -elements,' and in succession, impact the energy landscape, leading to further changes and the development of new energy products. Following the objectual senses, this cyclical process ensures that the established 2nd-energy-belt remains dynamic and constantly evolving.

(Again, following the conventional understanding of living beings, streamlining the above topics, it expresses and even emphasizes the timeless cycles of creation and destruction of objects. However, energy itself is eternal, and its diversity spans a vast range, from the microscopic to the immense within the matured state of the 2nd-energy-belt. The recognition of this diversity depends on the discernment capacity of individual beings, collective groups, or other intelligent species.)

→In the 2nd-energy-belt and from the standpoint of energy, the expression about the interlinking connections among diverse forms of energy and their products can be expressed equally to the point of the relationship between objects and their sub-products in the physical world. Just like objects can be parted into smaller components or give rise to new products to make new kinds of energy products with diversified dignities, each with unique properties and functions usually distinguished by intelligent beings based on the choice of some conditions.

But, in terms of energy and whatever their volumetric discernment, the **essence** of energy elements *(that is, at* *energy levels working components)* from an earlier state of the object will only be shared to small fields if there is no external imposition instantaneous upon dissevering the actual object, regardless of the object's size or volume.

It implies the previous example following a bottle filled with distilled, filtered, or purified water. Based on the standard interpretation, each drop of water in the bottle will occupy the **essence** as purified water.

(It is to remember for conceptual transparency from earlier illustrations, asymmetric sequences resulted in diversified techniques and implementation, which play a crucial role in creating 'energy dynamism' within the progressing Universe or IEfu. These orderings lead to the development of differential values, which further contribute to the diversification and evolution of energy products, noticeable in their diverse forms as objects in the matured 2nd-energy-belt.)

→In essence, the recurrent relation between energies and the created energy products in the matured 2nd-energy-belt emanates /generates and maintains a **certain level** of dynamism and a continuous process of transformation, diversification, and interaction, driven by the **inherent** properties and interchanges of the energies themselves. It differs from observable objects interacting in a void space, following the principles of 'action and reaction'.

*(Here, **'certain level'** indicates a parametric limitation, originating from the point of measurement to the limits of accessibility by measurer/s, extendable to an infinite range of barricading boundaries that secured the containing diversity of the matured 2nd-energy-belt. But, for the progressing fecund Universe or IEfu, that **'certain level'** can be demonstrated in the area **coverage** for preparation of the 'IEfu' within the Supreme - Universe. For this purpose and reasonability to understand the **coverage** of IEfu, I may refer to the drawing SU10 in this book.*

*Also, an objectual or the standard expression '**inherent**' indicates a characteristic attribute or existence in something as permanent. But, from an energy standpoint, the sense of '**inherent**' may be considered to have originated with the beginning of the actual Universe. And over time of advancement, it has become a practice and property of energy.)*

2. "Some quick clarification on the formation of the energy-atmosphere."
For ease of comprehension, let's assume that (E) represents unit-value energy rather than referring specifically to the term 2bE. This latter term indicates unit-value energies within the 2nd-energy-belt, which are in a state of equilibrium under certain conditions and are combined with composite energy elements.

Based on the objectually presented word-sense magnitude and fontal expression, the alphabets **E, E, E**, and so on have no difference. But in the mathematical sense, they are individual, and they can compute by the various number signs like plus, minus, times, division, etc.,
Somewhat may present like: -

- ► **"E+E+E+.... ∞ = "1E+1E+1E+.... ∞ = 3E...∞".**

Or, like**, no-sign E E E, that is,**

- ►**three times E** = **E3**

Above all, the plus signs and **no signs** fulfill the gaps concerning the **expression** for three individual units of energy as **E.**

*(Here, **no sign** means virtually no indication that something else exists is True or the possibility of transpiring.)*

From the standpoint of energy, this may say that certain distinctive classes of energy, as indicated by using the above number and numerical sign, fill the gaps of differential mono-status units of energies like **"E...+, -, ÷** and many, possibly to **∞"** that co-ordinated them under the conduciveness of **energy-atmosphere**.

i. "The development of the energy-atmosphere in pre-embryonic regions during the initial phases of the Universe's formation, encompassing several critical aspects of importance."

For conceptional clearance, it may recollect from earlier books that the Supreme Universe has the true eternal feature as 'no beginning and no end' status containing an infinite extent of Pure-energies at their Absoluteness-status, which means **any imaginable small unit of Pure-energy will occupy the same qualities that constitute the Supreme - Universe.**

*[{(**Some analytical clarification:***
*The point quoted as **'any imaginable small unit of Pure-energy will occupy the same qualities that constitute the Supreme - Universe.**' It is somewhat reasonable to exemplify that '**oxygen**' is understood by intelligent beings as a chemical element consisting of eight protons, eight neutrons, and eight electrons. Which is the fundamental scientific understanding of '**oxygen,**' regardless of whether it's measured in large or small quantities using standard methods for analysis.*

In the context of the Supreme Universe and the integrated vast expanse of 'Pure energies,' if we conceptualize a unit of 'oxygen' as analogous to a unit of 'Pure energy' at an "Absoluteness-status," then the components of oxygen like protons, neutrons, and electrons, can be justified as its working essences. When viewed from a broader perspective, these components of an oxygen atom might be analogous to the 'energy elements' that constitute a unit of

Pure energy. However, unlike the distinct and identifiable subatomic particles of oxygen, 'energy elements' of Pure energy are seamlessly integrated and indistinguishable from others.

In short, the above perspective implies that while physical entities like oxygen have discernible components, Pure energy exists as a holistic and indivisible whole.)

(One significant consideration is that Pure energies in the Supreme Universe cannot be conceptualized or justified as discrete unit/s to the same extent that living beings typically understand physical entities or substances. Therefore, the idea implies that recognizing the Pure energies as "units" at this phase is an assumptive status rather than a literal one.)

*→{(In this context, the ideologically designated **'motion element'** is given precedence due to the universally acknowledged understanding that 'motion,' which is always unique in many aspects from its lowest or highest levels of neutral and regulative roles compared to the sense of **complete stillness or cessation**.*

*So, based on mythical conception, the 'motion' or some form of dynamism may be considered a kind of celestial inherence to **'motion elements**.' Thus, it is logically acceptable why the common expression 'energy' applies to understand the many terms like shakti, power, fuel, active, lively, etc.)}*

*{(By the by, from the standpoint of energy of the matured 2nd-energy-belt, **'complete stillness** or **cessation'** masks like a '0' or its absolute position (**Ø**), which nourishes with some form of satisfaction for the 'brain processor' of living beings. Otherwise, on a technical aspect, 'energy' is eternally vital due to its ingrained 'motion element.'*

Further, in the context of the expression ***complete 'cessation*** *or motionless' analytically can speak, it is somewhat a satisfactory selection when the natural way evolves intellectual and non-natural creation measures to solve some purposes following some dimensional discretions. Something like an intelligent being/s, when measuring something through the brain processor, his active body mass, is conceptualized as a consolidated 0-value to get a satisfactory result for him or someone else.*

Similarly, imagine a non-natural intelligence capable of simultaneously accessing and analyzing multi-dimensional roles from various directions. However, from the stance of energy, the ability of such intelligence to measure would be constrained by the fundamental two extreme parametric boundaries, like a self-bearing baseline of zero-absolute value (Ø), and a far-reaching infinite value (∞).

To provide further clarity and realization on the above point, the diagram of the SU38 illustrates how both natural and non-natural intelligence conveniently exploits many non-requisitive entities regardless of various functional aspects that may act on the system, such as the continuous traversal of neutral and fast-moving energy, and unanalysable quicker actings items relying on their specificities or changing physical structures.

I suppose everyone is familiar with the purpose of various organs that evolved in the physical system of a living being, especially the eyes and the head, where the brain processor is encased and secure, primarily to perform the regulative roles of the physique.

Likewise, a living intelligent being, the invented non-natural intelligence, performs based on their constituent component used to design and program by the natural way evolved living beings mainly solve some purposes feasible to them, meaning both the inventor and invented items.

(Indeed, from the standpoint of energy, both the living intelligent beings and the invented non-natural intelligence are constraints in many aspects.)

"Following a mythological outlook, I imagine that an advanced generation of 'non-natural intelligence' (AI) is designed with many eyes, or bionic eyes, to provide a God-like, all-encompassing view. In addition to this holistic perspective, it would possess a highly advanced neural network, or something akin to a brain, with a proportional physical design that enables it to handle the vast amount of information received from its multiple eyes. This processing would occur instantaneously, at any fraction of time, far exceeding the standard analyzing methods in naturally evolved beings."

From an energy perspective, any enhancements to natural and non-natural intelligence will inevitably show signs of inadequacy when aiming for extensive energy coverage or the matured 2nd energy belt.

→The drawing SU38 includes marked points to aid understanding.

► The brain processor, or something like that, represents a zero-absolute value (Ø) for total molecular union or the accountable components or the overall mass-energy frame as non-natural intelligence.

► "Ranges of vision from an individual eye, varying from its closest to possibly farthest point."

► Many eyes have the potential to access more information than the natural human eye, which can only analyze a limited area at a time.

► A probable far-reaching infinite value (∞)

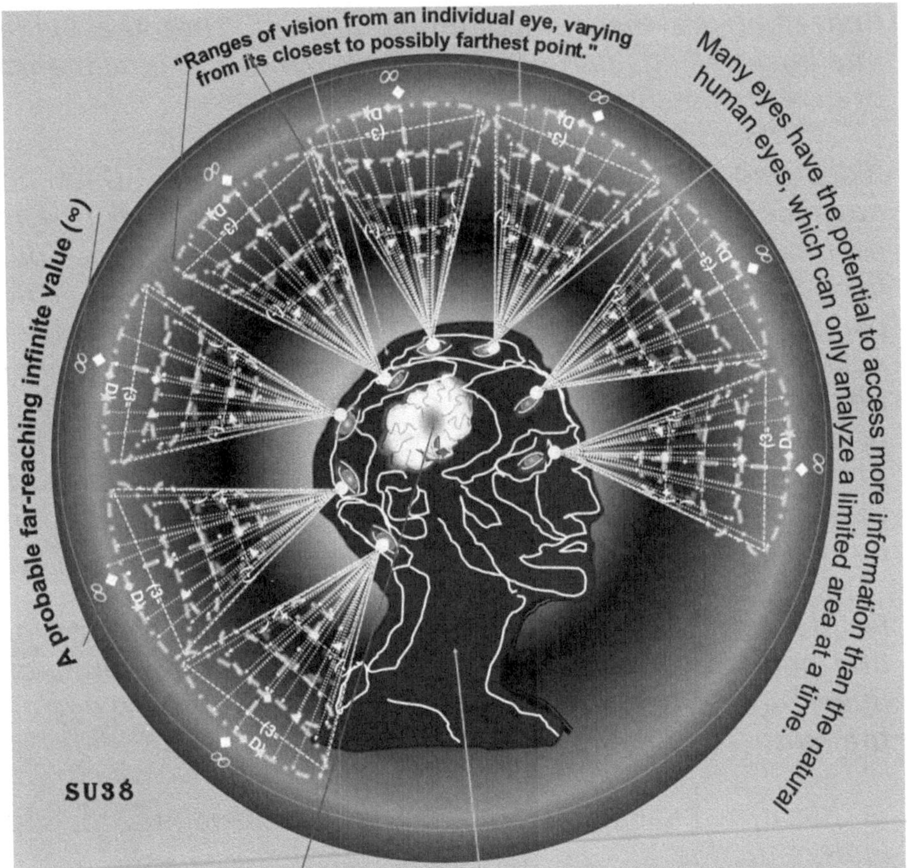

The brain processor, or something like that, represents a zero-absolute value (Ø) for total molecular union or the accountable components or the overall mass-energy frame as non-natural intelligence.

Besides, from the perspective of energy and its in-depth working essences, that is, energy elements, one significant point to remember is that whenever some amount of energy gives an accountable mass-energy frame regardless of physically recognizing living and non-living frame of reference. From their first point of ***appearance,*** *constituent working essence keeps performing by deriving internal physics, maintaining consistent transformation roles of energy, whatever energy levels loss and gain are performing in conducive and non-conducive surroundings.*

*(Here, the expression '**appearance**' technically may justify from a micro status unit or their amalgamated large frame.)*

Therefore, from the standpoint of energy and the natural progression of technological advancements, True measurements remain distant from reality when considering the vastness of the 2nd-energy-belt.)}

(*By the way, I have clarified the diverse roles of the 'motion element' in the relevant sections of my earlier books.*)}]

→Based on the standard conception of intelligent beings, it can be articulated and speak simple ways that an inconsequential, lowest-level '**energy atmosphere**' was initiated at the very primary stage before the gestation of the baby Universe began. This initiation occurred due to the unique and solo roles played by the deposition of motional values of motion elements at a specific zone of Pure-energies in the Supreme Universe *(I remember this topic illustrated in the 1st part of this book series).*

During this gestation phase, the asymmetric sliding activity of vulnerable Pure-energies in the prospective embryonic regions or IEsu of the Supreme Universe was responsible for preparing the environment or '**energy-atmosphere'** and some energy levels of **physics** that contributed with essential support for forming and advancing the gestation period of the baby-Universe.

*(This energy level's **physics** may speak as 'energy-values' when justifying as a measurable unit, or overall 'elemental value's' functions if one can think in-depth about 'energy elements' of that phase's energy level's vulnerable Pure-energy units.)*

However, the time-value's componential effectiveness on energy -Times, **energy-atmosphere** comprehensibly retained meshed among the units of energies within the

IEfu. It is due to the prevailing 'state-of-Absoluteness' of the Supreme - Universe.

So, the actual influencing energy zone in the fecund-Universe **(IEfu) 'energy atmosphere'** was much less effective than the **'energy–atmospheres'** of the 2nd-energy-belt. Due to this minimal extent of effective energy-atmosphere, from the initiation of the Universe, it has progressed exceptionally and inconceivably slower maneuvering process.

ii. There is no effective energy atmosphere in the matured 1st-energy-belt and 3rd-energy-belt.
In discussing the maturity of the Universe, with its **unique** three energy belts, the energies of the 1st and 3rd energy belts lost their compatibility in terms of energy atmospheres beyond the individual territories.

*{{(**Some analytical clarification**;*
*Philosophically, it may suggest that this **unique**ness stems primarily from the varying degrees of **quality** and **essence** inherent in the constituent 'energy components' across different energy belts.*

*In this context, when making three-dimensional assessments, it's important to note that living beings often use similar terms to differentiate between nearby and faraway things. These terms may include **quality**, quantity, **essence**, characteristic, and other comparable points.*

However, these terms may not always be suitable for evaluating and analyzing energy circumstances, which could provoke debate. It is because, for intelligent species in this habitable biological Globe, the brain's pre-set processing capabilities are delimited in many ways to decode near and far-reaching energy-level events and their outcomes.

Something like a standard, high-quality daily-use product might be deemed flawless by modern civilization, based on pre-set values either learned through external influences or evolved naturally. However, from an energy perspective, these justifications represent only a form of parametric satisfaction. Because energies perform severe momentary and accordingly, the proprietorial aspects of the occupying 'mass-energy' of the flawless daily-use product/s or unit/s keep changing regardless of any objectually determining parametric stance.

*"Therefore, in this context, the technical distinction between **'reality' and 'illusion'** depends on one's nature of conclusion and method of recognition, whether based on standard temporal perception or rapidly fluctuating energy levels performing their roles momentarily."*

Also, based on the Pure energies of the Supreme Universe, occupying energies by the three individual energy belts of the matured Universe are far from comparison to the specific way evolved living beings.)

*{(➢ In the context of the above points, **"reality and illusion,"** I want to place my ideas in a deep sense beyond the particle physics that may or may not be perceivable and acceptable through inbuilt delimited anchored sensory of bio-intelligent systems, which already explained in the book series 'Energy and The Complete Universe."*

➢ Anatomically, living beings independently perform as a union of multiple molecules. Groups of molecules construct different organs to give structural significance as a unit-value super-being.

➢ Each healthy molecule occupies some class values or classes of energy to perform with a well-order active life.

So, while molecules are in action, some form of signals or the same may title as 'energy signals' or 'e-signals,' which keep on releasing to perform Internal communication for their existence as a state of the stable molecule. It means molecules intercept others' e-signals of surroundings, process them, and then release them as expression /communication, like talks and gestures to express /communicate to other living beings.

➢ *Therefore, from the viewpoint of energy, it may justify that each molecule is conscious with some form of sensory that may be unalike to match or compare as a unit-value super being's conventional acknowledgment. All the e-signals that cover outputs of fleeting /momentaneous works of molecules somehow and someway relayed to the brain box, which acts as a final judiciary system to analyze and accept to derive action as reasonable for internal and external purposes of the physique.*

➢*Also, while the energy signals or e-signals with a particular spectrum travel to reach the Central processor* (brain box) *during the pass over a course, it mutates or relinquishes originality.*

➢ *Now, the word-sense illusion is considered a sensation that occurs in a person's mind and is expressed in the guise of delusion or false perception, which occurs when the brain processor misinterprets information due to natural or non-natural stimuli.*

➢ *Physical reality, as experienced by living beings, primarily analyzes three-dimensional accessibility determined by the* ***brain processor*** *to its surrounding accessible items in a time-bound manner, which also may mean any other ascertainable dimensional coordinate is determined based on the own built-in biological clock or acting living mobile clock.*

➢ *For computational and analytical purposes, the* ***brain processor*** *can be considered to represent a zero-absolute value (Ø) in terms of the total molecular union or the* *operating components accountable for the overall mass-energy system of a naturally evolved intelligent being.*

Or, applying a little philosophical sense, the above point may clarify that the brain processor unpretentiously tends to play a neutral, 0/zero, or non-partisan administrative role for the overall physique, whatsoever the internally deriving physics delivers to the processor due to the impact of surrounding natural or non-natural influences.

Something may be analog to a non-partisan Central Processing Unit (CPU) or brain of a computer, which administers following the instructions delivered according to the computer programs to process data, store it, and produce results.

➢ *So, the physical reality, or 'mass-energy' as a frame (mc2), can be ascertained when getting an appropriate structural significance to determine through the delimiting sensory organs of human beings. Otherwise, the in-depth connected works of energy remain obscure, a mystery, or a dream that the constituent molecules tend to express through the occupied /constituent central processor* (brain box) *of a living being.*

Also, it may not be the word 'mystery' that is the issue but rather a deficiency in the energy levels, either constitutional or technical, of a living being.

➢ *Further, expression through the word sense "Space" and "Time" standardly can be distinguished independently. The word-sense space, area, or volumetric concept is effectively apposite to any particle or mass-energy frames with structural significance. But the Time indicates the measured or measurable period during which (object) an action, process, or condition exists or continues with events.*

Otherwise, from the viewpoint of energy in the actual matured Universe, nothing can be justifiable as volumetric essences and be regarded as 'free' in terms of space.

*But, beyond the conventional opinion, instead, some form of energy occupies every nook and cranny in the matured 2nd-energy-belt in which mass-energy frame (mc2), particle, or object/s can play /move freely based on their "**volumetrically occupying space.**"*

Something, like the biologically evolved living species, is suited by nature, character, or design to a particular use, purpose, or situation, which became standard to fit and struggle based on their volumetrically occupying space in this biological globe.

*(Here, "**volumetrically occupying space**" refers to how an object or substance has a measurable volume, meaning it extends in length, width, and height and physically occupies a defined portion of space. In simpler terms, an object has a presence or "bulk" in the physical world, filling a specific area within the space around it.)*

- *The vision of countable Time can be determined when there is a reliable measure, such as a wristwatch, electric clock, electronic clock, or nuclear clock.*

Considering an object or objectual product to give /generate thrust needs some form of energy, either derived manually, mechanically, electrically, electronic application, etc. So, it is to speak that 'energy' itself is the root cause for deriving Time.

- *From the viewpoint of energy, straightly, it may think of a particle /object represented as a mass-energy (mc2) frame. Each unit-value mass-energy frame must constitute some form of energy components or the same considered 'energy elements' with independent working properties like pressure-element to create pressure, temperature-element to make temperature, motion-element to generate motion, and likewise numerous.*

So, at the in-depth energy levels, the 'motion element' is responsible for deriving thrust for commencing 'Time.' Accordingly, energy-level reality and the objectually determining illusion will remain conflicting topics to any physical status of a living being.

➢ *Considering the physique or a bio-physical structure to define the exactness of someone's illusion may be puzzling because out of 5 constituent sensing organs (those are the eyes, ears, tongue, nose, and skin), mainly and judiciously decide whether something is real or fantasy regardless, it is fed to the brainbox directly or indirectly in natural or artificially.*

➢ *Also, most people do not believe in illusion because their faith restrains within the individual temporal framework to a native perception and the trust that gradually grows through interactions in surroundings.*

Beyond that, anything intercepted by the deeply set molecules and processed through the brain box or thalamus remains unknown and unacceptable through the observable judgments, bearing in mind that perceivable areas /spaces and authentic Times of dreams remain unmatched.)}]

→Furthermore, following energy levels' quality and essences, the energy atmospheres of the 1st and 3rd-energy belts did not support the formation of energy masses, which will be suitable for creating next-generation energy products, such as objects and particulate matter commonly recognized by intelligent beings in the matured 2nd-energy-belt.

The presence of 'energy-atmospheres' and energy levels physics that supported instituting energy masses and the next generation's mass-energy frames as objects and other energy products contributed to the evolution of the dynamic structure within the matured 2nd energy belt.

Accordingly, philosophically speaking, some exclusiveness of atmospheres and their different levels that give rise to the physics for initiating and forging the micro level's objects of the matured 2nd-energy-belt are mainly confined within it, enclosed by the matured 1st and 3rd-energy belts of the Universe.

Therefore, the physics within the specific associations of atmospheres and the emergence of new, old, and advanced energy products are governed by certain functionalities and confined within the sanctuary of the matured 2nd energy belt.

In this regard, it emphasizes that any form of unit, whether in the pre-stage mass-energy frame or as a post-stage object, is formulaically associated and governed by the essence of the fundamental equation E = mc^2.

→Upon the maturity of the 1st-energy-belt and the 3rd-energy-belt, the effectiveness of their time values, particularly the 'energy times,' is restrained within their respective energy atmospheres.

It occurred in the individual 1st-energy-belt since the unit-value energies are meshed with others, influenced by the prevailing **near-absolute** feature. *(This "**near-absolute**ness" can be understood as a slight disintegration or deviation from the 'Absoluteness-status' of Pure-energy units.)*

But, for the 3rd-energy-belt, the effectiveness of time values reached the 'final absoluteness,' considered the finest form of energy or Purest- energies that institute the 3rd-energy-belt, situated far-end from the peripheral boundary or central location within the matured Universe.

Therefore, the time values effectiveness is determined by the two individual energy belts (*1st and 3rd-energy-belt*) themself, ensuring everlasting balance while the 2nd-energy-belt remains consistently active for internal development.

→Further, for the energies in the 3rd-energy-belt, there is no pinpoint to distinguish following the expression like the distinctiveness of its one type of component that is the finest form of motion elements or 'Purest-energy,' which is a **True singular** position in all aspects of energies. Due to this **True singular** status, the 3rd-energy-belt keeps no room to generate any differentiable atmosphere and physics, maintained as an everlasting benign **regulator** for the matured Universe.

*(Regarding the expressions, '**True-singular** and True-singularity' this has already been discussed in the book series "Energy and the Complete Universe.")*

In this context, the expression **'regulator'** may be exemplified by following a 3D assessment. Based on the consistently revolving yet non-counting regulator 'spindle' of a standard time clock, one could conveniently say that an 'observer' and the 'spindle' act as constants when analyzing the directional movement of the clock's hands to determine time. *(Indeed, many other components of a standard time clock participated in controlling the spindle and rotating hands, yet those remain unaccountable while something computes on standard times.)*

Therefore, the 3rd-energy-belt is appropriate to consider as an absolute constant and **regulator** for the matured Universe. *(In continuation of this topic, I plan to provide further clarification as relevant in the following parts of the series.)*

iii. Some points about the atmosphere and the distinguishing two commonized features like "Twig-atmosphere and energy-atmosphere."

In this mature 2nd-energy-belt, the atmosphere is filled up typically with micro levels of matter-element, matter-energy, matter-product, and so on, the ever-recycling ways appearing known and unknown miscellaneous progenies. Or, all of them may be dignified as objects, objectual-products, their spin-offs, and various yields, so applying the objectual sense, one of the types of atmospheres may commonize by being titled the **'Twig-atmospheres.'**

But in the growing *(mean before and primary stages initiation of forming objectual mass and their frame as mass-energy units)* 2nd-energy-belt, besides the **'Twig-atmosphere,'** there possessed a massive amount of energy, including some **free-flowing** energies like mixed-status, the rare presence ostensive form of 'Pure - energy' and rising ostensive 'Purest energy' that ply consistently and diligently.

Therefore, the atmosphere contains those sorts of mixed-status, ostensive of Pure, ostensive-Purest, and many other forms of energies *(but certainly not the later stages' fully grown 2nd-energy-belt and its emerging progenetic objectual products)*, which may commonize with another title that is, **energy-atmosphere** to identify as standardly acknowledging **'atmosphere'** in this matured 2nd-energy-belt.

So, in short, the matured 2nd-energy-belt, the standard way of recognizing the **'atmosphere,'** is a unison status of two introductory concepts as **'Twig-atmosphere'** + **'energy-atmosphere.'**

*(In the context of the expression **'free-flowing'** and other traditionally embedded terms used throughout the book series, these words may not fully capture the intricacies of*

energy level activities. They are employed merely as convenient tools to aid understanding, aligning with concepts that evolved from classical interpretations to suit the comprehension of modern intelligent beings.)

→Based on the objectually accessible atmospheres, understanding the features of the 'energy-atmospheres' and the standard physics within individual energy belts, such as the 1st, 2nd, and 3rd energy belts, remains distant and contradictory when approached analogously through the lens of individual living beings and their collective existence as a species.

It is because conclusions are drawn from the available energy essences, which serve as a referable framework, whether considering an individual living being or the unified intelligence of the collective group.

Also, it is to speak that the consistent transformation practices throughout the matured 2nd-energy-belt reserved the rights of modification and aptitude of containing energy from its primal stage.

From the standpoint of 'energy times' in the independent energy belts of the Universe, features of the 'energy-atmosphere' are remote to comprehend and conflict with the common ideology about the standard ways of recognizing atmosphere, which stance as a twig status of combining 'Twig-atmosphere plus energy-atmosphere,' mainly analyze based on observably objectual properties and the objectually deriving times or standard clocks' times.

From the standpoint of 'energy times' within the independent energy belts of the Universe, the multiple attributions of 'energy atmospheres' are challenging to comprehend and often conflict with conventional ideologies about standard methods of recognizing atmospheres.

This discrepancy arises because energy atmospheres are often analyzed based on objectual properties and the confinement of timeframes, which scrutinize and follow the recognition of classical objects, such as the measurement computed based on the standard time clocks. This approach serves as a simplified "twig" interpretation of energy dynamics rather than capturing the more intricate and complex nature of the 'energy-atmosphere' itself.

3. 'A little note about the sublimation/transition period of 2nd-energy-belt's 'energy elements' that led to the formation of various objects in the matured 2nd energy belt.'

→ ("Some points to consider regarding the manifestation of object-oriented energy.")

From the early primal stages, 2b-energy-elements *(2bee),* or units of 'energy elements' *(here, instead of fixed unit-value energy, I conveniently refer to their momentarily fluctuating constituent operating essences, that is, energy elements),* have undergone countless cycles of shuffling and reshuffling within the 2nd-energy-belt over eons of energy time. As a result, these later stages developed 2b-energy-elements ***(2bee),*** which eventually contributed to the appearance of the primary phase essential functioning values for the fabrication of objects, besides the emergence of components for objectual energies.

In addition, during this tumultuous period, the growing 2nd-energy-belt also generated or gave rise to other perplexing and non-objectual energies due to consistent evolutions of physics derived at energy levels or from elementary levels of energy elements. The combined efforts of both prospective objectual energies and existing non-objectual energies subsequently led to the creation of diverse progenies, such as objectual items.

In this biological nature, the commonly recognized atmosphere and its composite components are primarily viewed as the surrounding environmental processes and their pervasive effects, which are fundamentally the roles of descendants of 2b-energy-elements (2bee) or advanced generations of primordial energies of the 2nd-energy-belt, unlike some mysteriously convinced observable conclusions.

By the way, it is justifiable that, regardless of recognition based on the structural significance of contained energies, whether as the smallest element, a particle of matter, or the largest structure of a material product, all genetically imbued with the essence or features of 'Pure energy.' It may also imply that any unit-value energy in the matured 2nd-energy-belt tends to reach a state of balance, which is distant but may justify, similar to the feature of 'Absoluteness-status' of unit-value Pure energy.

i. "Based on the observable practices of living beings in this adaptable world, the properties of individual energy units within the matured 2nd-energy-belt, regardless of their forms, cannot be accurately quantified numerically, nor can their magnitudes be considered equal."

In the context of the unit-value energies, the terms of measurement and functionality of any forms of mass-energy frame in the matured 2nd-energy-belt, their performing magnitudes cannot be accurately quantified numerically, nor can they be considered equal, based on the standard practice of living beings to comprehend assessable items for classical applications.

(This point also emphasizes the innermost significance of the equation E = mc^2 beyond the aggregating observable essences.)

It is due to the **value differences** *(or differences in value)* that arise during the encapsulation of **'completely bodied**,' or possibly better expressed as **'concentrated'** energy elements. These energy levels' fundamental elements ultimately form unit-value mono-state energy, or their union as 'mass-energy,' exhibiting versatile properties under the conduciveness or reasonability of atmospheric conditions.

*{(Beyond the contemporary context that standardly applies classically to comprehend objects and objectual items, technically, at energy levels, the above point as "**completely bodied** or possibly better expressed as **'concentrated'** energy elements" may be elucidated a little more. It suggests from earlier books and speaks that if it is intellectually countable or terminable some amount of energy or naturally positioned as mass-energy or a mass-energy frame by using some improved application, following the constancy as infinite (∞) for the vast energy - Ocean in the matured 2nd-energy-belt.*

It proposes that energy elements, elemental values, or elementary-level physics values behave fleetingly within any mass-energy frame. They consistently shift roles from a tenable state of energy, energy elements, and elemental values, generating energy physics that strives to achieve zero-status stability. It occurs by aligning with the external or atmospheric energy physics surrounding the unit-value 'mass energy.')

(Here, from the standpoint of energy, one point to remember is that the pre-and post-stage mass-energy frames that represent observable and non-observable objects are far distinctive in disposition, unlike the classically applied expression 'equal,' which means as an average upshot, not the energy levels momentary impact.)}

Here, one issue is to point out that the expression 'constancy' objectually refers to the quality of being unchanging, stable, or reliable over time. In a scientific context, it could refer to certain quantities that remain unchanged under specific conditions, like the speed of light in a vacuum. Overall, constancy emphasizes consistency and reliability in behavior or characteristics.

However, at energy levels, there is no true constancy. Technically, this suggests that from the tiniest possible units to larger ones, all at their unit state or mono entity tend to seek a balanced status within the conduciveness of the energy atmosphere in response to the surrounding physics. Besides, at energy level times are the integrated values derived due to most momentary roles of motion elements.

Besides, one point to remember is that the ostensive Purest-energies constitute the life-like soul of the matured 2nd-energy-belt and are available in every nook and cranny of the matured 2nd-energy-belt, representing an ever-lasting constant or base for everything due to its bewildering highest kineticism and neutrality.

Here, on the aspect of 3-D computation and analysis, at energy levels, the following point may justify the possibility of fundamental three-layer **constancies** to comprehend the comparable functionalities of unit-value mass energy and mass-energy frames in the 2nd-energy-belt.

Those **constancies** are the; -

►'mightiest peripheries at one end that checks the 1st-energy-belt, and the other is the 3rd-energy-belt' of the matured 2nd-energy-belt.

►The soul or the most dynamic and neutral state, 'ostensive Purest- energy.'

►And at energy levels, the ever-versatile but objectually presumable **'zero-status stable** unit-value mass energy and mass-energy frame.'

*{(Here, the expression "**zero-status stable** unit-value mass energy" could imply a foundational state that underpins all energy forms that has the potential to turn into a state of unit-value to represent mass-energy, mass-energy frame, and thereon consequences of elemental values, energy value or speak as 'physics values' administer them to an advanced progenies object.)*

(In this context, it is crucial to recognize that the deeper meaning of the expression 'motion' is an essential concept across various fields, such as philosophy, science, mythology, sociology, and likewise. It addresses the criteria of these disciplines as they relate to living beings, who, through their intellectual capacities, experience and appreciate motion while existing in a habitable environment, even from a state of relative stability or equilibrium or zero-status-stable position.)

These dynamics are somehow comparable in how mathematical **numbers, symbols,** and arithmetic **formulas** define their values and relationships cooperatively, maintaining some level of obscure constant that represents an overall base for satisfaction.

*[{(**Some analytical clarification:***
*In the context of the expression, '**value differences**' refer to the 'diversified proprietorial aspect' of energy levels that are somewhat analogous to the 'analysis and characterization' of objectual items based on some degree or level of incompatible attributes, using comparable strategies.*

*One important thing to remember, as defined earlier in the context of purified water and oxygen, is that 'energy is energy' {(like the '**E = mc^2**,' philosophically may mean '**E**'* *energy maintained its* ***equivalent essence*** *(marked by '=' sign) to the* ***'mc'*** *or unit-value mass-form energy whatever their distinguishing discretion based on the indicated symbol lambda* ***(^)*** *and numeric terms (like* ***2****)}.*

Further to quote, regardless of any distinct recognition based on observable senses, every conceivable or even minuscule unit-value energy embodied with some reasonable essence of 'Pure Energies,' like the feature of state-of-Absoluteness that motivates any unit or units of their mass-energy form or object to achieve some level of state-of-stableness, which have carried on genetically from the Great-grandmother, the Supreme - Universe.)

*(Here, the expression "****equivalent essence****" refers to the core or fundamental nature of two or more things being the same or having equal value or significance, even if their forms or appearances differ. In other words, it describes things that share an underlying or intrinsic quality despite external variations.*

For example, if two different systems or objects are said to have an "equivalent essence," they operate on the same fundamental principles or exhibit the same essential properties, even though they may look or function differently.)}]

→Furthermore, those **numbers, symbols,** alphabets, and similar representations may remain within certain limits of magnitude, making them adhere to specific levels of typical understanding or patterns of resemblance on comparable strategy to living beings.

For example, numbers like '1, 2, 3, 4...9' and symbols such as +, -, and ×, etc., each of those possess individual values and limitations defined by their respective scales or magnitudes.

However, when repeated as 1, 1, 1 or 2, 2, 2, or symbols like +, +, + or -, -, -, they appear alike or exhibit resemblance.

Yet, when those numbers come close or combine with their common or uncommon one, 'with or without' attaching any symbol, their merged form compels a change in their ingrained features like one and one makes 11 (*eleven*), nine and one makes 91 (*ninety-one*), etc.

→So, remotely, based on the 'energy time' of an 'energy clock,' the above can briefly elucidate and link to every momentum value of a motion element within a unit-value energy. It, in turn, generates a **differential value** from any point of constancy confined by the energy atmosphere and other loss and gain maintaining cooperating tools.

In essence, if those individual integers, symbols, or their combined mixed status presume to represent a specific unit or a unit-value mass-energy frame constitutionally integrated with distinguishable energy elements, then each momentary shift in the Motional value of the motion element *(that is, an essential component for deriving motion)* drives a difference, thereby compelling a change from the earlier constancy determinable based on **manipulating events.**

(Here, ***manipulating events*** *means 'influencing or controlling circumstances to achieve a desired outcome, which can involve strategic planning, persuasion, or even deception to steer events in a specific direction. Depending on the context, it can have neutral, positive, or negative connotations. What specific context is someone judging to achieve some result?').*

Since motion elements are inherent in perpetual motion, thus every momentum value to any form of unit-value mass-form energy represents an exchanger involving some form of loss or gain from an assumed constancy. As a result, the status of any unit-value energy continually shifts from any point of presumptive fixedness, which is unmeasurable and incomprehensible, following any standard 'time gap' determined by the intelligent living beings in this habitable globe.

In this context, the understanding of energy by intelligent beings across diverse environments and evolutionary paths, limited by the rules of composition or design, along with the unique types of energies that constitute each individual being, can be regarded as a reference for mass-energy frameworks.

Consequently, the magnitude, numerical value, and distant characteristics of energies are limited by their specific way of design and further influenced by the involvement of mediating entities, whether they are living or non-living intelligent products like (AI), or advanced progenies.

ii. "Note in short about the 'neutral-energy,' energy-elements, effective elemental-value, and absolute-equilibrium state."

The following are the definitions of a few technical terms related to performers of in-depth energy-level roles as mentioned in the above subject.

►**Neutral-Energy:** Refers to an energy state that does not impact other energies or systems 'straightly.' It acts as a balancing element, much like zero in a numerical system, providing a neutral reference point.

Further, the expression 'neutral' represents somewhat a baseline that neither adds nor subtracts from the system, which is similar to the role of a balancing point and analogous to how the symbol zero operates in mathematics.

►**'Energy elements:'** are the individual and most fundamental components or units that constitute unit-value energy or make up a more prominent energy system. Each 'energy element' has its 'own value and role,' and together, they build up the comprehensive energy of the system, characterized as a mass-energy frame.

'Energy elements' can be distantly compared to numbers in a numerical system, each representing a unique quantity or value, and mathematically can be manipulated to some level of satisfaction.

►**Effective 'Elemental-Value':** This is an active or impactful value of an energy element when contributing to the overall system. 'Elemental value' can be considered the productive role each energy element plays. Or, elemental value is possible to justify as the degree to which an energy element affects the overall balance or state of energy, somewhat similar to how the position of a digit in a number gives it its place value *(e.g., ones, tens, hundreds).*

►**Absolute-Equilibrium State:** It is a state where all energy elements are perfectly balanced, with no net movement or imbalance to a system under the conduciveness of surroundings.

Also, it represents the ultimate state of stability or neutrality, where all energy elements have reached equilibrium or stabilized, and there is no driving outcome for further change, akin to achieving a complete and total balance in a mathematical equation.

{(From an energy perspective, while the above point may seem similar to a general idea of equilibrium, it suggests that any mass-energy unit or framework tends to maintain a comprehensive balance that accounts for consistent loss and gain within the context of its surrounding environment.

It contrasts with a parametrically determined object-oriented approach. Instead, at the energy level, the conduciveness of the atmosphere, defined as the interaction between the "Twig" and the "energy atmosphere," determines the ideal state of equilibrium for a given mass-energy unit or framework, maintaining interlinking relation with the vastness roles of energy-Ocean in the 2nd-energy-belt.)

Further, for more clarity, it emphasizes that rather than objectually defining achieving a static balance, mass-energy units exist within a dynamic framework influenced by their environment. It suggests that the interactions between mass-energy units and their surrounding energy atmospheres play fundamental roles in maintaining equilibrium.

In this view, the "Twig + energy atmosphere" concept implies that the properties of the surrounding environment, including energy exchanges and interactions, significantly affect the state of the mass-energy units, mass-energy frameworks, or objects. This dynamic interplay challenges traditional notions of equilibrium that might focus solely on static parameters. Instead, it emphasizes a more holistic understanding of where energy flow and environmental factors are central to the behavior and stability of mass-energy systems.)}

→Further, to illustrate the above concepts, following the standard practice of living beings, it can be justified that the '0' symbolically represents the neutral in placement for the other natural-born ascendant or descendant arrangement of the numeration systems *(1 to 9).* In this analogy, energy elements *(somewhat like those digits)* can shift or change in value, retaining the 'neutral energy' *(0-like characteristic)* as a constant, ensuring that movements stay within a balanced range of an overall system between its lower and upper limits.

Again, when all the integers directionally move following the natural base '0' in order from 1 to 9 and possibly more. That is simply 0, 1, 2, 3, 4……9……, ∞, where the zero /0 and ∞ can consider their absolute status (*can be realizable from the drawing SU39).*

But when the numeric integers bi-directionally *(back-and-forth)* move based on neutral-value-zero and the limits of infinite. Then, the zero /0 and '∞' represent their common position, maintaining the absolute or unmodifiable value for any directional deviation. Or, say, a satisfactory 0-integer and an unknown deal ∞' act as a state of objectiveness for the standard arrays of directional divergence for other numeric integers.

Like the bidirectional deviation of integers with positive and negative number symbols keeping '0' as a neutral position *(indicated in the drawing SU39).*

∞...-9, -8, -7, -6, -5, -4, -3, -2, -1,
< ‘0’ >
+1, +2, +3, +4, +5, +6, +7, +8, +9...∞

However, in the binary numbering system, the two possible independent values of the digits 0 and 1 are typically interpreted as representing two distinct states. The digit 0 is generally associated with the 'low' state, indicating absence, off, or false, while the digit '1' is linked to the 'high' state, symbolizing presence, on, or True. These two values form the foundation of binary logic, allowing complex information and processes to be encoded and computed in digital systems by utilizing combinations of these fundamental high and low states.

But, for energies, those objectually determining bidirectional topics can be extended to many directions following a common deviation point, as indicated in the drawing SU39.

→It assumes that if all numeral systems express a unity or unison state of composite energy, then for a unit-value 'mass-energy,' the numbers from '0 to 9' would represent different types of integrated 'energy elements' and distinctiveness on the aspects of values and proprietorial requirements.

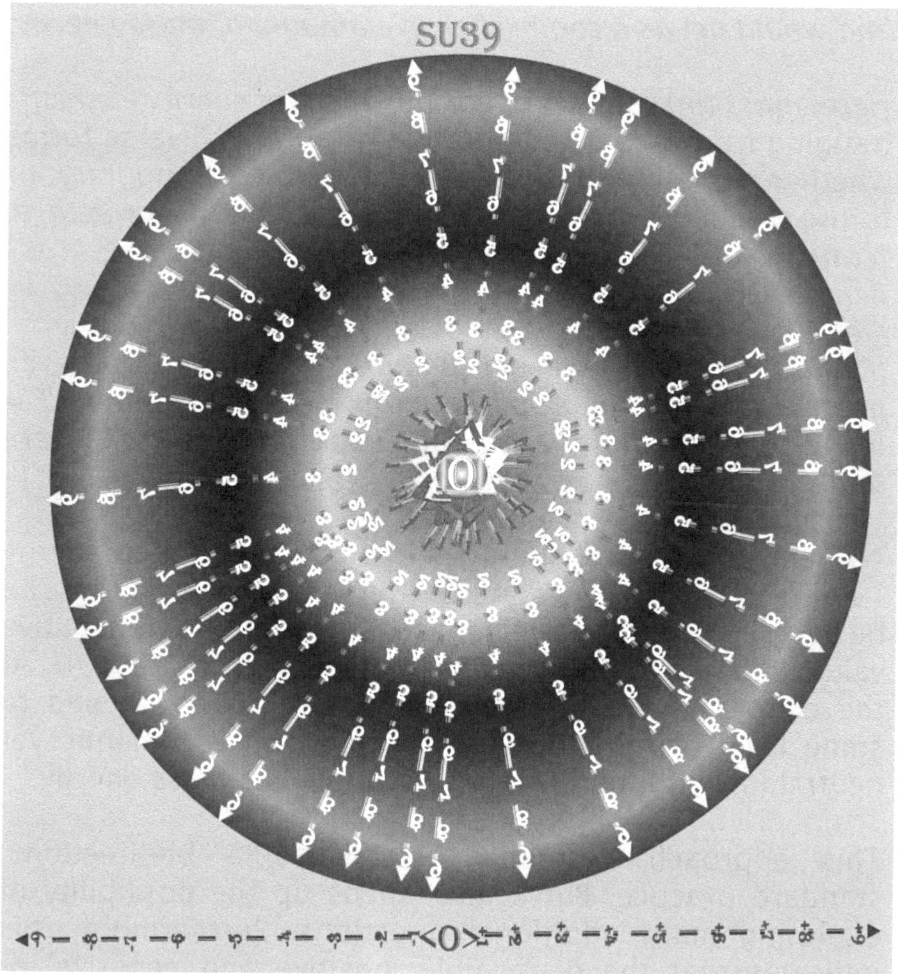

A reasonable conclusion from the drawing 'SU39' is that if the intuitiveness of its overall importance represents a mass-energy frame, then from the base value or point '0,' all directionally dispersed, digitally sectioned, and sub-sectioned rays or lines *(whether representing right and left,*

positive and negative, or other micro-level directional deviations with specific values) will indicate differences in their functional usefulness.

{(Analogously of E=mc^2, for further clarity on the topic, if we assume that the alphabet 'E' represents a unit of mass-energy frame or (mc^2)*, then constitutionally representing 'mc' would act as a comprehensive amount of mass-energy.*

Here, mass (m) consists of various types of energy elements (ranging from 1 to 9 and potentially more, in diverse forms whether represented by using historical, contemporary, or future symbols*), including the motion element* 'c' *or neutral featuring '0' or like 'soul' all within the specific scope denoted by '^2'.)}*

(Regarding the noteworthy dynamic, regulatory, and even neutral roles of the proposed letter 'c' in the equation 'E = mc^2,' explained in the previous book series as and where appropriate.)

→Again, for the 2nd-energy-belt, all the integrated energy - elements within any state of unit-value mass-energy can be assessed comparatively, based on positive and negative values relative to a neutral base point, such as zero /0. At the elementary level, this numeric '0' can be likened to some level of motion element (c), which is dynamic yet neutral in attribution within any unit-value 'mass energy.'

This approach not only establishes a bi-directional standard practice. But it also opens up the possibility of exploring many additional directions. Furthermore, this indicates that the discernible 'positive and negative' or 'upper and lower' degrees of energy levels can be evaluated as appropriate to analyze energy level functionalities.

Additionally, the potential for all attainable energy elements in any unit-value mass-energy or referable frame can be constrained to specific limits for numeration purposes, adhering to the standard low and high states of binary numbering systems utilized by intelligent beings.

(One important point to remember is that energy performance does not follow a specific direction like the clear discernment of observable items. Instead, energy functions multi-directionally, adhering to consistent elementary levels 'elemental values' or distantly similar to physically regulating 'physics values' activities.

Therefore, for energy level activities, some particularity of directional measurements through the physical systems of intellectual beings will remain a comparable strategy, retaining some hidden undertaking as 0 or neutral value for computational perfectness.)

→It may provide further clarity by incorporating philosophical ideas, assuming that any complex numbering system formulated with multiple individual small units and conditions reflects the typical attribution of energy. Alternatively, it is possible to simplify by defining each numbering unit as part of a composite frame representing a comprehensive mass-energy framework like the drawing SU39.

So, in this context, the digits from 1 to 9 represent various individual statuses as mass-energy frames *(or countable objects*), integrated with assortments of energy elements which may be indicated by numbers and multiple symbols as appropriated like the contemporarily denoted different decimal levels of positive (+), negative (-), and so on, probable of distinguishing based on the importance of their applications *(for convenience of grasping the above topic, it may depict with a drawing SU40, which explained in one of the earlier book series).*

For clarity and ease of understanding of intelligent beings, it may consider an example of the mono state of 1, 1, 1, or

2, 2, 2, where the integers appear identical. However, for mathematical purposes, they hold self-sustaining status.

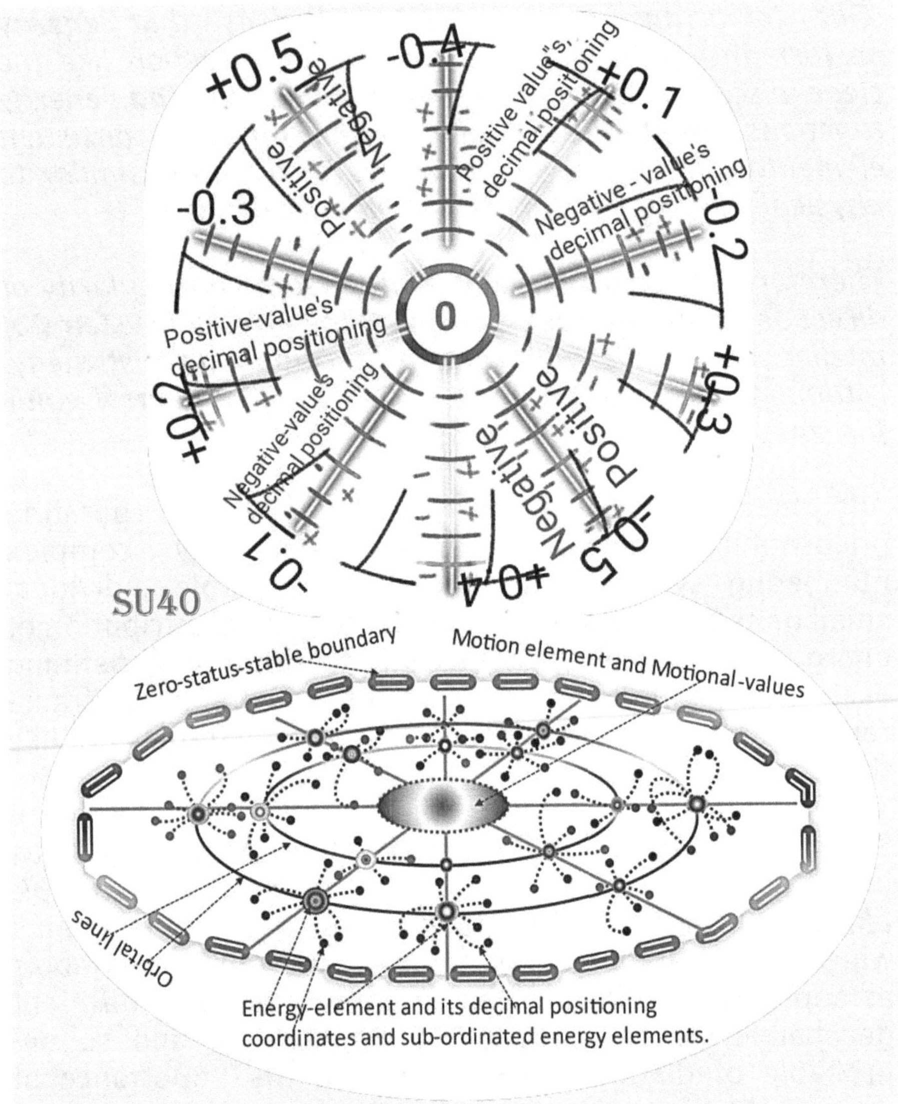

But, when those integers combine with another similar or distinct one in terms of mathematics and also consider unit-value energy, the resulting values or outcomes will not be the same.

It also suggests that energy possesses the fastest and most fleeting characteristics compared to any micro-level unit of mass-energy, mass-energy frame, object, or particle, as it is constantly engaged in securing its parametric boundaries in coordination with the surrounding energy physics.

Further, analogously from an isolated viewpoint of an assumable unit/s of Pure energy, which exists at their 'Absoluteness-status' in the Supreme - Universe, the afflicting unit-value energy or their advanced mass-state appearance in the Universe tends to secure parametric boundaries of their individual, which is a type of energy level congenital leverage.

(By the way, I recall this point discussed in the series of books "Energy and The Complete Universe," which I will elaborate further in upcoming volumes as appropriate. One key point to note is that, at the energy level of any of the Universes in this matured 2nd-energy-belt, everything has a direct or indirect causal relationship, technically comprehended as an expression of genetic leverage or influence.)

{(Plainly, we should not forget that energy (represented by the alphabetic symbol 'E' in the equation 'E = mc^2') *has vast implications beyond the discernible limits of living beings. Its influence extends throughout the 2nd-energy-belt, impacting realms that surpass the familiar roles shaped by evolution within this biological globe.)}*

→In the 2nd-energy-belt, the expression evaluations or devaluation of energy elements in any unit-value energy can also be justified comparably, using the standard positive and negative values' derivement based on some state of dynamical but neutral featuring base point, such as zero /0 or (c).

Indeed, analogously, the roles of energy elements within any energy level unit are relatively faster than the comprehensive outcome of advanced stages frame, like a large object, micro-level objects, or microparticles.

Since 'energy' is intrinsically dynamic, ranging from the lowest degree of potential to its peak kinetic state, even though it ensures regulating the **discipline of Universal laws of physics** in the vastness energy-Ocean of the matured 2nd-energy-belt. Besides,

Technically, in explaining the reasoning behind the perfect **discipline of the Universal laws of physics**, it can be pointed out that the expression 'causal relationship,' or cause and effect, refers to how one variable (*the cause*) directly or indirectly leads to changes in other variables *(the effect).* This concept plays a vital role in determining the eternality of the mature, proprietorial characteristics of the 2nd-energy-belt.

Furthermore, aggregating all attainable energy elements within a possible standardized unit-value energy could demonstrate a 'mass-energy' frame, and their functional roles are conceivably analog based on the low and high values in binary systems, with the potential for many more variations realizable following the advancement of intellect.

→Also, if there is any potentiality to become a separated energy unit or mass-energy frame (*denoted by mc^2*) occupying some form of known, mysterious, and non-analogous energy elements along with their working essences or elemental value when considered as whole or comprehensively. The unit-value mass form energy should contain a neutral-energy status (*such as 'c' or motion element)* encapsulated with the other energy elements and affectable elemental values.

So, mathematically, it is considered that during a state of equilibrium, any unit-value energy, and even the smallest of its solitary form as a unit-value particle or object, must contain some class of neutral energy.

It also clarifies the core concept of **neutral energy** (*denoted as 'c')* being present in any standard form of unit/s of energy or an advanced stage's comprehensive values 'mass-energy frame,' which is ready to regulate neutrally besides the constitutionally set dynamic roles of energy components or energy elements, whatever it depicts as a discernible or indiscernible object following the standard acceptance as static or some form of dynamism.

Also, when preferred for representing the elemental nature and values of energy elements, this can be justified by the expression 'utmost infinitesimal.' If 'e' denotes an energy element, then any elemental value (ev) acts as its coefficient. It implies that when 'e' represents a specific type of element, 'ev' conveys the magnitude or quality of that element. These definitions of 'e' and 'ev' can be manipulated following the standard mathematical models used by modern intelligent beings.

{(I remember that the technical topics of energy elements and elemental values are illustrated in appropriate points in almost all the book series, as these concepts introduce entirely new perspectives for advanced generations and processes at the most fundamental elementary levels of mass-energy, mass-form energy units, or the prospective objects.)

(Further, ideologically neutral energy refers to a status of unreactive energy while maintaining some specific valuable properties, limits, or narrow interactions with specific or non-specific energy elements within a unit-value mass-energy system.

'Neutral energy' can be compared to motion elements when representing a bulk state within the mass-energy framework. Alternatively, 'neutral energy' could be viewed as analogous to the concept of a spiritual soul in mythological contexts, symbolizing the essence of a living entity. Because the fundamental characteristics of motion elements, neutral energy, and the spiritual 'soul' share similar features, such as representing a core or essential quality.)}

iii. Remembering the point mentioned in the earlier book, the Supreme - Universe is the mother of all genesis, Part 1, where it indicated that, based on the observable sense, technically, it is possible to justify that the Supreme - Universe and the Universe act like

$$'0^{SU} + 0^{U} = 0^{Superbly\text{-}embeded},'$$

unchanging the originality of individual proprietorial values.

(Based on the standard mathematical conception of two '0's, this may imply that the Supreme - Universe and the Universe's comprehensive feature remains at a state-of-Absoluteness. Alternatively, in another sense, objectually may be interpreted with some satisfactory expression like 'absolute nothingness' or, in a more non-technical sense, as a 'miraculous assignment.')

Thus, following the standard mathematical conception of contemporary living beings, in a very remote manner on the aspect of energy elements (*sign* e^{s}), it is convincible that the Supreme - Universe represents similar to a number figure, '0,' and its feature is analyzable as:

0 = (neutral-energy) = (e + ev^{0}) = (energy-element + elemental-value^{0}) = (Inert/unreactive along with the coefficient).

→ *(Instead of this intricate analysis, it is better to simplify the above terms as 'state-of-Absoluteness.')*

Here, for the Supreme - Universe, its working essence serves as a neutral position since there are no functionally active elemental values, or based on the objectual sense, the same somewhat may speak like the:

"effectless 'elemental-value' = ev^0 = 0"

(Based on the above clarifications, in the Universe, some form of bulk energy can be featured as neutral energy since their acts are comparably effectless, like 0 /zero. Technically, those forms of neutral energy are the thin /fine form of extractions during tumultuous activities of energies in the hugeness 'energy-Ocean' of the Universe.)

So, on the point of effectivity, it considered that the Supreme - Universe itself stands as actual **0000**…**.......**n, **∞** or say **zero$^\infty$**, or the sense-like effectless coverage far beyond as 'no end and no beginning' standing.

Accordingly, it may elucidate that the Supreme - Universe itself is at the absolute-equilibrium state, or speak, at an unconditional state-of-Absoluteness, or may symbolize as '**0absp**.'

→However, when considering the energies of the Universe, this also simplifies the idea that any state of unit-value energy, or an energy element with a coefficient of ideological value zero (0), exhibits an equilibrium state under certain conditions.

Additionally, it clarifies that the objectual representation of the value 0 (*zero*) does not signify 'absoluteness.' Instead, from an energy perspective, it reflects certain objectual limitations relative to a native standard. These limitations may correspond to logical values or degrees that provide solutions to specific needs for living beings, particularly within a biological context.

The expression {(*energy-element*) + (*elemental-value*)} altogether may symbolize 'eev,' is an ever persistence, or say, integrated to every unit-value energy, whatsoever a comprehensive manner as a union of unit-value energies, or a unit-value-energy, mass-energy, mass-energy frame or a prospective object in a compressed state with its 'energy elements.'

It also may mean that the constitutional physics or elementary levels' physics of the unit-value mass-energy or mass-energy frame reached some 'state-of-stability' under specific conditionalities of the surrounding atmospheric physics.

iv. "Brief discussion on the aspects of 'evolution' and 'devaluation' about the integral components as energy elements from their earlier states in unit-value energy."

In simple terms, 'evolution' refers to the development or progression of something, and 'devaluation' means a reduction in the worth or importance of something, often viewed in materialistic terms as a loss of value relative to others.

From an energy perspective, **integrally constituent components** mean the fundamental components or the essential elements at the energy levels undergo continuous change. Both the status of the energy element and its value shift as the identity, disposition, or characteristics of the recognized energy element evolve from their previous state.

(This most momentarily changing roles of energy elements in various forms of 'energy-mass' have been explained differently throughout the earlier books, whenever relevant.)

{(The phrase ***'integrally constituent components'*** *refers to functional parts or elements that are essential and fundamental to the whole. In other words, these components (*such as energy elements*) are integral to the structure or function of the entire system. They cannot be completely isolated without affecting the integrity or operation of the system based on its predefined or recognizable status. Of course, the changes are consistent, which can be expressed using words like 'evolution and devaluation' from an earlier status.*

It is important to note that the roles of energy are often momentary, unlike the more stable and parametrically recognizable objects or an observable system.)}

In some ways, this connects with our intellectual understanding of 'life and existence' within the framework of biological nature. To clarify further, the instantaneous estimation of the comprehensive energy value, based on the prior state of an energy element, relies on specific parameters. As a result, the function and significance of the energy element evolve depending on how its elemental value is applied.

In other words, as energy is measured and utilized in various contexts, the role and importance of the energy element change accordingly. So, this makes it evident that the traditional concept of the comprehensive outcome of any unit of mass energy practically originates or derives from its integrally constituted energy components.

→In this broader context, it can be insightful to compare and understand different energy belts, *such as the 1st, 2nd, and 3rd energy belts of the matured Universe,* specifically someone when realizes how intelligent species interact within their physical world, represented as 'mass-energy frameworks' amidst the vast dynamics of the 'energy -Ocean.'

(Although traditional views of living beings may seem disconnected from that perspective, from an energy standpoint, the functions of energy remain technically True, just as a conscious being might live a calm, balanced life, unaware or uncared for turbulent energies nearby or at vast distances. However, energy constantly surrounds and influences everything from an unimaginably micro to a massive discernible structure.

Similarly, one might think of oceanic fish, living their entire lives underwater in the vast ocean, never needing to become aware of or understand the complexities of energy levels that function around them, just as many living beings remain unconcerned with the deeper realities of energy.)

It may indicate, based on the earlier mentioned topics and considering the effectiveness of objects, one could infer minimally that the overall capacity of energy levels' energy element in the 1st-energy-belt is correspondingly extreme than the energy levels, which is possible to uncover in the developed 2nd-energy-belt.

Furthermore, this may suggest that objects are only ascertainable within the 2nd-energy-belt. As a result, intelligent species evolve to become adept at analyzing objects and observable phenomena.

However, based on the 3D judgments of intelligent beings, the energies in the matured 2nd energy belt are more disruptive compared to the high-capacity but less-disturbing energies of those residing in the 1st energy belt.

(Of course, there are many more terms that analogously can be employed to analyze the quality and capacity of energies within the 1st, 2nd, and 3rd energy belts, which I will elaborate on in upcoming books as appropriate, and from technical stances on the 'construction and

development' of the actual Universe will be explored systematically in the new book series.)

v. This point focuses on the functional limitations of energy elements and how exceeding the specific parameters can direct and disrupt balance.

A discussion on the limitations of the workable range of elemental values of energy elements may highlight the significance of consequences and their limitations.

Much like numeric elements, where values other than zero (0) or the base position have specific functions, the elemental value of an energy element is similarly natural ways confined within a defined workable range.

Extending beyond this parametric span or range can lead to disturbances in normalcy or deviations from some standardness of an individual and consequential effects on the total arrangement of an established system.

For clarity and ease of understanding, drawings SU41, SU42, SU43 and SU44 illustrate this concept, showing how integers from 1 to 9 function within the individual ranges, starting from their point of origin. If someone considers an integer to represent an "energy element," then the range covered by that integer is analogous to the span of recognizable elemental values or functional limitation of the energy element.

Similarly, the rules governing numeric scaling ensure that each digit maintains its distinct boundary to preserve precision. In the same way, each energy element occupies a specific range from its origin, shaped by the conduciveness or standardness of its surrounding environment.

→Further, to sharpen the distinction between mono-directional and multi-directional perspectives while linking and persuading the idea to someone who realizes works in-depth on energy.

Herewith, based on the limitations of conventionally acknowledged numeric elements used by bio-intelligent species on this Earth, the depicted drawing SU41 suits as an example.

Part (a) in the drawing indicates a mono-directional perspective, where one may conclude that each number belongs to a numeral group that evolves systematically in ascending order or a mono-directional discernment.

Part (b) in the drawing SU41 indicates that part (a) is a mono-directional or (+) discernment, which can be devalued in the reverse direction by employing some perceptible sign like negative (-) or something else that is suitable for distinction.

So, part (b) reasonably represents two directional earmarking from a common base point like 0, which means a 3D coverage since the positive (+), negative (-) direction, and the 0-base overall act as the standard three-dimensional discernment to a living being.

However, intellectually advanced species in this bio-nature could augment their thinking values and intellectuality to some extra directions suggested in part (c) of drawing SU41.

Similarly, part (d) of drawing SU41 hints at a multi-directional perspective, where for analysis, someone of bio-living species could maintain himself as a base or reference point for simplifying 3D concepts and some reasonable extensions to extra dimensions.

Drawing SU41 helps illustrate how each numerical element and its associated values possess its individuality. It makes it easier to grasp how the 'energy levels' can be evaluated from multiple perspectives beyond the limitations of specific observable concepts.

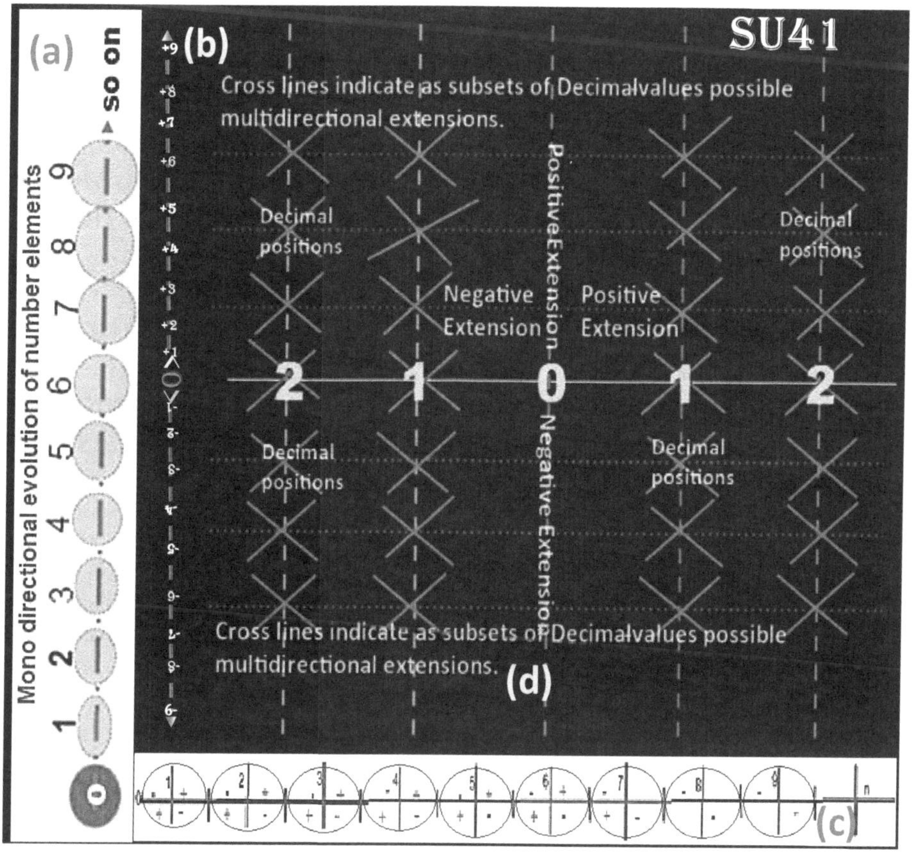

Furthermore, the drawing employs simple directional lines and accompanying discernible positive and negative signs to indicate movement and value, irrespective of whether these signs represent an original or extended decimal position.

→Indeed, from the perspective of energy physics, human physical stability entirely relies on the perfectness of tackling the consistent internal (means, *bodily representing bio-logically evolved mass-energy framework)* and external (means, *atmospheric, that is, Twig + energy atmosphere*) multi-directional functionalities.

(In the context of energy levels, this concept of 'multi-directionality' and multi-dimensionally is illustrated in one of the book series "Energy and The Complete Universe.")

However, the brain processor can analyze some averaging 3D limitations, leaving plenty of non-essential things as non-assessable for prolonged. Hence, cooperation as a group form becomes essential to earn multiple knowledge that derives for consistent energy levels' physics, acting on everything naturally.

Concerning the 3D point of limitation, I may include a drawing (SU42) from an earlier book series to ease understanding, which may give basic ideas concerning how the physical system sensibly analyses the accessible surroundings.

Arrows in the drawing SU42 indicate the following points: -

►i. From the perspective of constituent molecules of a unit-value lively being, brain power tends to work multi-directionally to an infinite limit of an individual living body mass or referenceable frame of 'mass-energy.'

►ii. Brain Processor and the standard Judicial Centre for the whole-body mass of a representing living being.

►iii. However, due to structural delimitations with sensing faculties like the eyes, observably discernibility is typically restrained within the perimeter of three-dimension, from an assumed center line or axis that acts as a measurable constant for the total body mass of the living unit that constitutionally works together or union of multiple molecules.

►iv. From some dimensional projection, an overall body mass acts towards the assortment of physics values roles, which covers the entire influential areas as a parcel of a sack of energy, representing a unit-value lively human being.

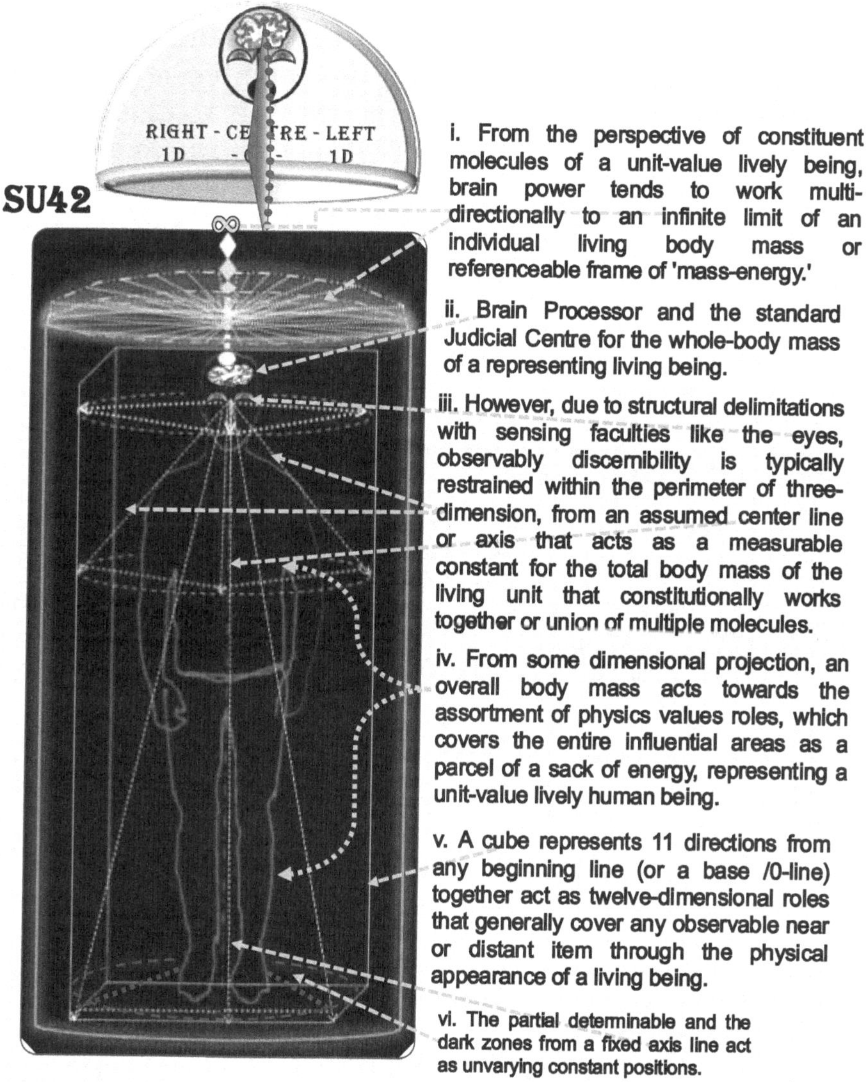

►v. A cube represents 11 directions from any beginning line (or a base /0-line) together act as twelve-dimensional roles that generally cover any observable near or distant item through the physical appearance of a living being.

►vi. The partial determinable and the dark zones from a fixed axis line act as unvarying constant positions.

vi. It may elucidate the distinction between conventional sense evaluation and its more complex energy-based multi-directional perspective.

Depicting with a drawing SU43, it attempts to demonstrate how bio-intellectual beings typically evaluate numeric elements directionally unless and until these evaluations tend to manipulate beyond the standard for personal and group satisfaction.

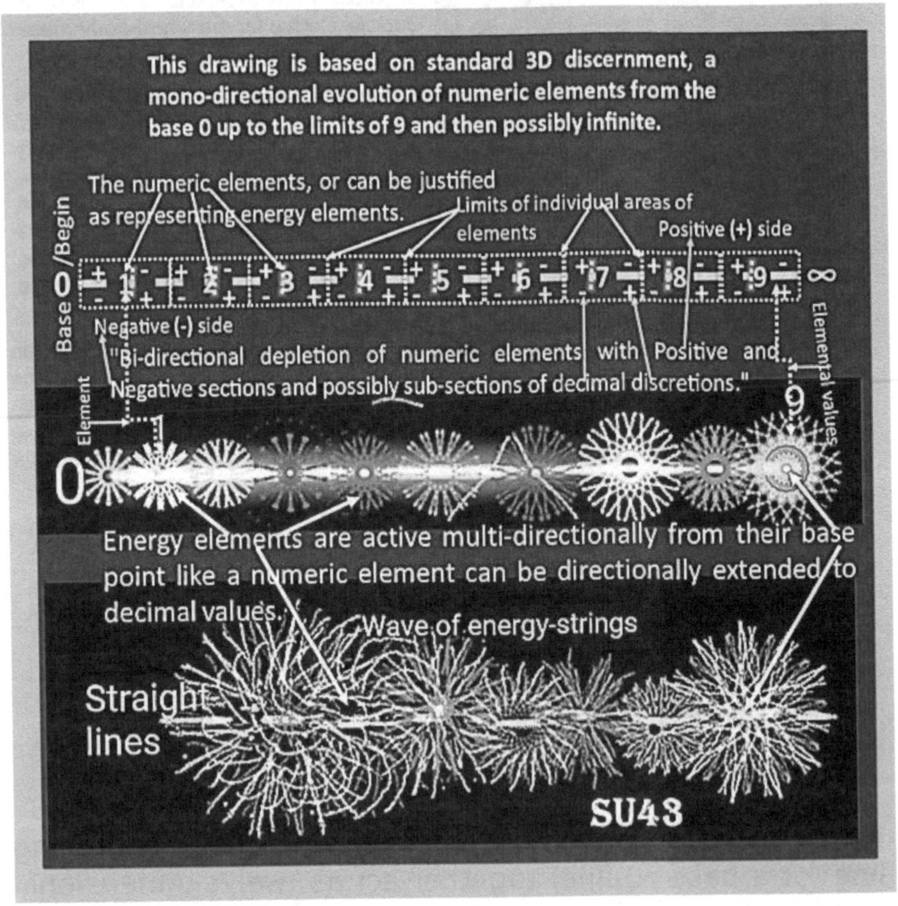

However, from an energy perspective, the numeric-elements can be comprehended and appraised in a multi-directional sense, anchored by a base position or center point, whether indicated or denoted by the number symbol '0' or any other individual alphabet like 'c.'

So, based on the arrangements of number figures, analytically in a multi-directional sense. Reasonably, it may think about the energies of the Universe and remotely analog the integrated multiple types of energy - elements. Though, the aspects of capacity or quality one can again extend to certain limits of its complete dilution upon exhaustion of elemental values at a state of individual dignity.

So, when analyzing the arrangement of numbers from the multi-directional viewpoint, one might draw parallels to the energies of the Universe and the various integrated energy elements. These elements extend until they reach the limits of their complete dilution, as their elemental values (*indicated in the drawing*) are exhausted and conclude from a state of proprietorial individuality.

Further, the above clarification effectively highlights the realism of the principle of "energy transformation," emphasizing that it occurs almost instantaneously at the fundamental level of "energy elements." This idea could be a core concept in understanding how energy transitions or shifts happen in the most basic units or components of 'energy' itself. It also captures the essence of a fundamental and dynamic process within energy systems.

→Also, drawing SU44 is a reasonable status to indicate how numeric elements from 1 to 9 in a ray or line can be utilized in complex multi-directional arrangements, considering them similar to energy elements with distinctive qualities like the degree-wise arrangement of numeric elements or numbers from 1 to 9.

Each ray, line, or string originates from a common base point like '0' or 'c,' then dispersed multi-directionally, maintaining level-wise discipline on individual orbital lines from the 1st to the 9th of dispersal. Therefore, all the same types of numbers are settled systematically from the 1st orbital lines to the 9th or end of orbital lines or end of some stipulated orbital lines.

In addition, a complex design is demonstrated in drawing SU44, which hints at how energy elements in a mass-energy frame function in a systematic order under a conduciveness of surroundings.

Indeed, from a standard viewpoint, if someone can justify that the complex elemental arrangement in drawing SU44 represents a mass-energy framework, then at the energy level, the framework will consistently be functionally capable of maintaining loss and gain ratios, balanced under that conduciveness of atmosphere or atmospheric physics.

The drawing SU44 again indicates two distinct sections: one showing "a simple way of elemental arrangement" and the other demonstrating "a limited extent of elemental complex arrangement."

Another drawing in SU44 elucidates how 'numeric elements,' or those similar to 'energy elements,' can be reasonably applied to understand positive and negative pathways in standard computational analysis.

Furthermore, the drawing highlights the "orbital arrangements from a common base point 0 or (c)."

In terms of energy levels and energy elements, (c) represents a source of motion alongside its neutral and regulatory characteristics. This source or origin of motion derives from the unceasing roles of the 'motion element' that helps maintain a systematic order of orbital lines and ensures the consistent functioning of energy elements in harmony with suitable environmental conditions.

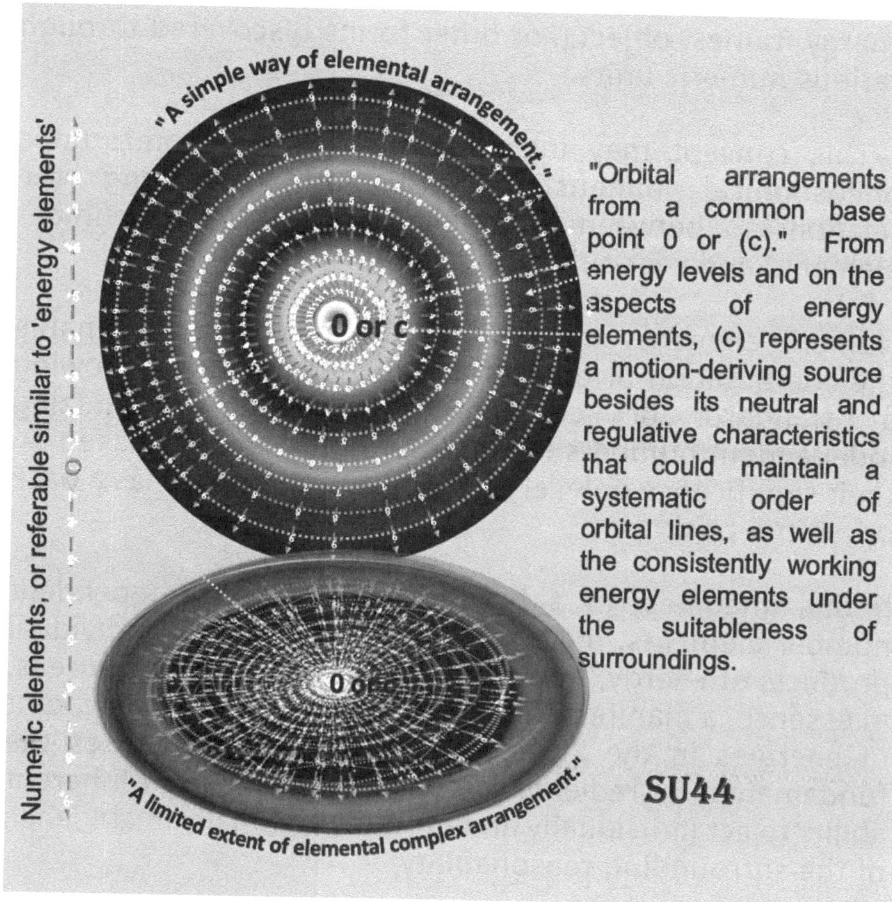

vii. Furthermore, I recall an earlier book series (*Energy and the Complete Universe)* where it elucidated the significant symbol '0' and its reasonable application for energy-related purposes. In the mature second energy belt, intellectual beings conveniently used the concept of zero ('0') as a foundational base for all other primary numeric elements or figures, ranging from 1 to 9, without combination.

→Similarly, in the context of energy and a facilitative condition, the motion element *(c)* acts as a foundational base point that regulates the other energy - element/s.

It *(c)* governs their *(energy elements*) state, whether all together represent unit-value energy, mass-energy, mass-energy frames, objects, or other forms discovered through various numeric units.

→This concept may introduce a slightly uncomfortable philosophical dimension, particularly regarding the relationship between bio-intellectually designed number elements and ever-subsisting energies.

Numeric elements are a natural gift discovered to enable the computation of objects and objective items, following a standard of comparative strategy. Organizing independent numbers in an evolutionary manner based on their significance or level of importance retains '0' as a base or neutral position.

If one imagines, in a broader sense, that independent number elements, including the symbol '0,' are individual products of energy, it suggests that each number figure is, in essence, a manifestation of energies. From the viewpoint of energies in the 2nd-energy-belt, energy becomes the fundamental ingredient of each number, with an inherent ability to act individually and multi-directionally in abidance of the surrounding reasonability.

From an energy perspective, analyzing the evolution of individual number figures becomes complex. Conceptually, all forms of energy are locked or stabilized in a "Zero-status-stable" position, a constant state that allows for the discrete and orderly emergence of numeric figures. This positioning serves as the foundation for the symbolic representation of numbers.

Conclusion: The Nature of Energy and Reality

I believe this book series will bridge the educational gaps for future generations, especially for those drawn to immersive virtual realities and fantasy worlds governed by the laws of material nature. Fantasy worlds, much like the intricate systems of living beings and biologically influenced habitable planets, mirror the coordination between energy and the materialistic surroundings we inhabit.

In fact, from the energy perspective, inhabitable nature, similar to this biological Globe, is governed by the interrelationships between energy levels. These relationships intertwine with physics, which, in this context, extends beyond traditional physical concepts. The vast 'energy ocean' connects visible and invisible products and forms. These (*objects or forms*) are the progeny of energies, directly or indirectly interconnected, regardless of how or where they appear within the matured 2nd-energy-belt.

(It's important to recognize that the term 'physics' is no longer limited to physical impressions. Instead, it has broader implications when considered at energy levels. Here, physics reflects the foundational principles of an energy system, representing a mass-energy framework that applies to both living and non-living forms correspondingly.)

It is time to recognize the wisdom passed down by our predecessors, who warned us about the deceptive nature of what we observe. They emphasized the concept of energy, which, though the 'energy' is often impalpable, is the factual driving force behind everything. Energy transcends the visible and invisible, existing beyond form, size, color, or any physical attribute, yet it is fundamental to the operation of life itself.

Generations of knowledge gathered through practical experience, abstract thinking, and celestial studies like science, philosophy, and mythology continue to propel humanity forward. This accumulated understanding from generations helps living beings evolve and lays the groundwork for future generations to thrive.

→Further, the earlier book series, titled Energy and The Complete Universe, published in four parts, has delved into the role of energy in evolution, which extends far beyond traditional concepts.

Indeed, energy has a broader implication for deriving life and guides evolutionary processes. Accordingly, the laws of universal and natural 'energy physics' influence the adaptation and development of life forms across different habitable environments, much like this biologically evolved globe.

Considering the gradual advancement of living species from the Primeval era, I believe that future generations of intelligent beings will surpass conventional ways of technological integration for augmentations. It will occur at their in-depth understanding of energy's nature, and its applications transcend the limitations of classical and particle physics, enabling the exploration of new habitable zones and the realities of different energy levels activities.

Indeed, a question may arise: Are energies themselves 'conscious,' like us?

In this regard, I wish to highlight that the creation of diverse linguistic elements, alphabets, words, and grammatically organized sentences did not emerge instantly. Instead, they have a deeper connection to energy levels within the human species.

This connection has enabled discoveries and new findings throughout human history. Changes in linguistic expression have occurred naturally over time, adapting to new purposes and facilitating the advancement of future generations by making it a simple and comfortable communication technique for all as one human species.

A human being is typically considered 'conscious' as long as they are healthy and supported by a conducive environment. From a technical perspective, this means that consciousness evolves in living beings as they experience life, interact with nature, and acquire knowledge directly and indirectly while struggling at the molecular levels by birth and existence.

The brain functionally represents a biological living processor, and the bodily materializing molecular components work in tandem to serve the overall consciousness of the lively physical system.

From the energy perspective and on the aspects of 'energy physics,' any form of unit-value energy, mass-energy, mass-energy frame, or objects can justify 'conscious' whenever they achieve a unison state upon appropriate application of suitable energy levels surroundings in a place of the vast energy-Ocean.

In these cases, energy elements momentarily alter their state, and once they reach unison or 'zero-status stability,' they exhibit consciousness by applying internal elementary physics. This process seeks to maintain balance and stability within the system, countering external forces.

So, the 'energy elements' are the most fundamental at energy levels, where consciousness appears to tackle the individual position of unison or zero-status-stable unit by applying internal physics that checks and attempts to balance itself as a unison or zero-status-stable unit from external applications.

In this way, even celestial bodies within the 2nd-energy-belt can be considered conscious. However, this concept may be challenging to grasp for biological beings since they evolved following the confining conditions of our habitable planet.

It is important to remember that energy, as a source of all life and vitality, carries its inherent dignity. Our predecessors recognized the eternal nature of energy, which they applied in many ways, including natural and unnatural compromises for good, ill, and impartial, as part of their survival, besides cultural and mythological understanding.

While modern science and philosophy have accumulated vast knowledge, future generations will likely delve even deeper into the realities of energy and its workings. This understanding will guide them toward becoming true intellectuals, capable of controlling individualism and egoism to prevent hassle-free conflictions and unnecessary destruction of our planet, the biological globe we call home that provides safe and sheltered space for living beings, besides authorized enjoying natural way evolved loving care with family and friends.

→Further, it is beneficial for Harnessing Energy Knowledge to Transform Aggression into Harmony.

While modern science and philosophy have made significant strides in understanding the Universe, the future promises even deeper insights into the nature of energy and its profound workings in shaping the living world.

As humanity continues to uncover the intricate relationship between 'energy and reality,' the profoundness of knowledge on 'energy' will play a pivotal role in shaping the intellectual growth of future generations.

A heightened understanding of the roles of energy in our existence will empower individuals to prevent destructive tendencies. It will encourage a shift from aggressive individualism and ego-driven behavior to a more harmonious collective approach, improving the pretentiousness of humanity.

With this enlightenment on wisdom at energy levels, people will learn to control the natural and unnatural ways of deriving impulses, which lead to conflict from some satisfactory understanding and call for unusual destruction that harms not only others but also disrupts directly or indirectly the delicate balance of our biological globe, the Earth that sustains all life.

This livable planet, after all, provides a sheltered and nurturing environment for living beings. We must protect this natural space, not only for ourselves but for the generations to come.

As knowledge of energy deepens, individuals will realize that True intellectualism lies not in domination or ego but in a balanced and mindful existence where the living species and the habitable planet, as a collective, will be promising.

The shift in consciousness will foster an atmosphere where conflict will lower to the least possible level, consequentially promoting cooperation with new-founds and the ability to control aggressive impulses.

Therefore, individuals can focus on nurturing relationships, finding joy in family and friends, and the simple pleasures that arise from a natural, evolved way of life.

So, by improving upon learning to harmonize our inner energies with the outer world, the customarily destructive revolution of ego-driven conflicts will ease up to a culture of care, understanding, and sustainable blossoming.

Constitutionally set energies of living beings and surrounding energy levels conditionalities emphasize the essentiality for intelligence and emotional growth, aligned with the natural laws of energy in a locale, to bring about a more peaceful and cooperative future while living under the care and nourishment of mother nature bio-globe.

I sincerely appreciate your interest in exploring the ideas and insights shared in "The Supreme – Universe is the Great-grandmother of All Genesis, Part 2."

I hope this journey enhances your understanding and curiosity.

This installment of the book series, titled "The Supreme – Universe is the Great-grandmother of All Genesis," offers detailed explanations complemented by numerous illustrations to enhance readers' understanding. As the first edition of this new series, it surpasses the depth and scope of "MYTHOLOGY, PHILOSOPHY, and SCIENCE: Book for the New Generations, Volume 1," which was first published on May 15, 2022, under a free KDP ISBN.

9 789334 140682

Printed at Repro India Ltd.